Quest for BIGFOOT

A Novel of Adventure for Young People

by

Kyra Petrovskaya Wayne

illustrations by
Maret Rehnby

ISBN 0-88839-396-2
Copyright © 1996 Kyra Petrovskaya Wayne

Cataloging in Publication Data
Wayne, Kyra Petrovskaya
Quest for Bigfoot
ISBN 0-88839-396-2

1. Sasquatch—Juvenile fiction. I. Rehnby, Maret. II. Title.
PZ7.W35129Qu 1996   j813'.54   C96-910421-9

All rights reserved. No part of this publication may be reproduced, stored in a retrieval system or transmitted, in any form or by any means, electronic, mechanical, photocopying, recording, or otherwise, without the prior written permission of Hancock House Publishers.
Printed in Canada

This is a work of fantasy. All characters and incidents, as well as the places, are entirely the product of the author's imagination and are used fictitiously. Any resemblance to actual locales or persons, living or dead, is purely coincidental.

Production: Gerald Pauler
Editing: Nancy Miller
Cover: Gerald Pauler

Published simultaneously in Canada and the United States by

HANCOCK HOUSE PUBLISHERS LTD.
19313 Zero Avenue, Surrey, B.C. V4P 1M7
(604) 538-1114   Fax (604) 538-2262

HANCOCK HOUSE PUBLISHERS
1431 Harrison Avenue, Blaine, WA 98230-5005
(604) 538-1114   Fax (604) 538-2262

Contents

I	The Dream	5
II	New Friends	17
III	Where's the Proof, Dr. Stone?	25
IV	Visit to the Nootkas	34
V	Potlatch	43
VI	Joey's in Trouble	50
VII	Archery Competition	62
VIII	Journey Back	69
IX	Dr. Stone's Lab	76
X	Chickie's Revenge	83
XI	Honor Bound	89
XII	In Search of the Expedition	97
XIII	Dr. Stone's Discovery	106
XIV	SASQUATCH!	118
XV	Visit with Dr. Stone	126
XVI	Raven's Cousin	134
XVII	Going Home	140

For my grandchildren Nicky, Chris, Emily, Natalie and Kyra. May your quest for discovery be forever strong.

Previous Publications by Kyra Petrovskaya Wayne

Kyra
Kyra's Secrets of Russian Cooking
Quest for the Golden Fleece
Shurik
The Awakening
The Witches of Barguzin
Rekindle the Dream
Max, the Dog that Refused to Die
L'il Ol' Charlie
Quest for Empire

I

The Dream

Steve stood at the Western Airlines terminal in the multi-level Seattle-Tacoma airport, waiting to be picked up. Weighted down with a camera, a sleeping bag and a suitcase stuffed with T-shirts, underwear and jeans, Steve was ready for the adventure of a lifetime.

"You must be Steven Bradley." Steve turned around.

A stocky young man of about twenty, dressed in a checkered lumberjack shirt and faded jeans stood smiling at him. His long, straight hair was held by a leather band around his head. His face was the color of copper and his dark, narrow eyes appraised Steve with friendly curiosity.

"I'm Jim Brown. The locals call me Raven's Wing. I am Joey's cousin. He's sorry that he could not come to meet you, but you'll see him later. Welcome to the Northwest!" They shook hands.

"Do you have any other luggage? I hope not. You seem to be really prepared," Jim smiled.

"No, this is all."

"Okay, let's shove off. We have quite a way to go." Jim picked up Steve's sleeping bag and headed toward the doors.

The parking lot was filled to capacity. "I am parked in a no stopping zone," Jim said. "I had to use ingenuity to park at all.

Just keep an eye open for traffic cops while I remove the distress signal." He led Steve to an old jeep parked at a curb solidly painted yellow. The hood of the car was lifted up and there was a white piece of paper torn from a notebook impaled on the radio antenna. Jim snatched the paper off and slammed the hood down.

"Climb in, friend, we're off!" Steve threw his suitcase in the back of the jeep and settled himself next to Jim, placing his camera on the floor at his feet. He felt shy in the presence of Jim, so handsome, so sure of himself. He sought desperately for something clever to say, but nothing came to his mind.

Jim watched him from the corner of his eye. He could tell that Steve was nervous.

"There are no billboards!" Steve suddenly exclaimed. He had never been north of Los Angeles and the difference in the landscape, so green and lush, so brilliant in comparison to the dusty-brown hills of California, was startling.

Jim chuckled. "Yup...some years ago my people ganged up on the state government and demanded the removal of all advertising from the highways. So, now—no more billboards. We went back to nature. We're going to take a ferry," he continued. "It's a short ride across the Puget Sound, but we'll have enough time to poke around the boat. I like to check out unusual places, don't you?"

"Yes, I do." Steve suddenly felt at ease.

At the ferry terminal a line of cars was already moving along the designated boundaries and into the boat's car deck. Jim fell in line and inched his jeep close to the car ahead until the bumpers touched. He set the emergency brake and they climbed out.

"Let's go on the upper deck. It stinks down here."

They climbed the narrow stairs emerging from the bowels of the ferry into bright sunshine and a brisk breeze smelling of sea and fish.

"That's better," Jim said. "I hate to be cooped up in some tight spot like down below."

"Me, too."

Puget Sound was alive with sailboats and small motor crafts. From the vantage point of the upper deck the motor boats looked like water bugs dashing about, crossing one another's wake. As the ferry began to move away from the landing, the tall modern buildings facing the sound suddenly seemed to burst out in flames as they reflected the sun, brilliant as gold.

The city grew smaller in the distance, as diminutive green islands came into the view. Steve watched the changing seascape, thinking that perhaps he shouldn't have left his camera in the jeep. He could have taken some great snapshots of the shoreline...

"Some view, eh? You should have brought your camera," Jim said, as if reading Steve's mind.

"Yeah," Steve said watching the broad wake of the ferry.

"I always promise myself to carry a camera along whenever I am not working, but I never do," Jim said.

"What do you do, Jim?"

"I'm a fisherman. In the old days I would have been a whaler, but we don't hunt whales anymore. And I also go to college. Someday I hope to go to law school. My secret ambition is to become a lawyer like my Uncle Joe. And what about you? What's your secret ambition?" He glanced at Steve sideways. "You have one, don't you?"

Steve felt flattered that Jim, a grownup, favored him with his confidence, but he hesitated. "You won't laugh?"

"Me? Never!"

"I want to be an explorer. I want to find Sasquatch."

"You...what?"

"I want to find Sasquatch," Steve repeated, blushing painfully, now fully expecting Jim to burst out laughing, despite his promise, like most adults have done whenever he had mentioned his secret ambition.

Jim remained serious, although a smile flashed across his eyes.

"Oh? You know Bigfoot's Indian name?"

"Sure. I know a lot about him!" Steve boasted, encouraged by Jim's interest.

"Well, pal, you've come to the right place," Jim grinned. "Our folks here claim that they have seen Sasquatch many times."

Steve felt the palms of his hands become sweaty, as they always did when he was excited.

Jim continued, "If you're lucky, you might come upon him. He's very selective, they say. He doesn't show himself to just anyone."

Steve sighed. "It would be *fantastic!*" he thought. "Me, finding Sasquatch!"

"Have *you* seen him?" he asked.

"Me? Never. Frankly, I don't believe he exists." Steve felt disappointed. He had hoped that Jim would guide him in his quest.

"But don't go by me," Jim said noticing Steve's disappointment. "At least half of our people believe that Sasquatch lives right next to us, in the forest."

The ferry reached the other side of Puget Sound and docked. Jim and Steve returned to the jeep.

"Here we go, into the wilderness of the old Indian country!" Jim said as he drove off the ferry and into the tree-covered hills. "And we didn't even explore the ferry!"

A magnificent panorama of green mountains and blue waters opened before Steve's eyes. With every turn of the winding road a new vista spread before him. There were mountains and deep canyons covered with thick forests and down below, the narrow crescent-shaped lake shimmered and sparkled, sapphire blue in some places, silvery in others. There were hardly any houses on the periphery of the lake. Steve had the strange feeling that there was not another human soul within a hundred mile radius.

As if to shatter this illusion, two logging trucks loaded with huge logs chained together, thundered by them, breaking the silence.

"We're almost there," Jim pointed to a thicket of woods.

The Makah Indian village where Steve was to spend the

eight weeks of his summer vacation was perched on a high point above the Pacific. Surrounded by thick forests of hemlock and spruce, the village consisted of two rows of houses on each side of the narrow macadam road, a brick church with a tall white steeple, a general store, a gas station and a narrow, one-story cement block building housing the community center. The houses were small and most of them in need of repair. Some were surrounded by small orchards of apple trees just shedding their spring finery of blossoms, the others stood forlornly, exposed to the ocean winds.

The village looked like thousands of others, nothing marking it as an *Indian* village, except for the canoes. They were everywhere. In front of every house there were at least one or two canoes, intricately carved and decorated with the tribal symbols. Some were propped on special supports, the others lying on the grass like beached porpoises.

"These are the whaling boats," Jim said pointing to a large canoe. "We, the Makahs, were famous along the coast as whalers," he continued driving slowly along the main street. "In the old days, there were no better whalers than the Makahs! My grandma could tell you about the old days if you're interested."

"I sure am," Steve said. "I've read a lot about Native Americans, but I've never met one."

Jim smiled. "Well, you're talking to one now."

They came to the end of the street. Tree branches stretching from one side of the road to the other, interlocked their arms, creating a vaulted, green ceiling over the darkened tunnel where the street became a road again, disappearing into the forest.

"See that cliff?" Jim pointed to his left. "Down below, on the beach, is where I live, in a trailer. But I'll take you to Uncle Joe's now. You'll be staying at his house."

He drove back towards the center of the village, stopping at a white frame house. The house was larger than the others, set in the middle of a neglected apple orchard and surrounded by an open porch. Three dogs were asleep in the shade of the porch, sprawled on the cool grass. A yellow cat perched on the railing, was licking her paw, passing it over her muzzle as if waving to Steve.

Above the front door, a large carving of a raven astride a whale greeted the visitors. "The Chief's ancestral crest," Steve thought.

The dogs, wagging their tails, lazily stretching, arching their backs and yawning, sniffed at Steve's legs and circled around his luggage. Satisfied with their inspection, they retreated under the porch. The cat totally ignored Steve, primping herself with her rose petal tongue. Jim opened the screen door. "Uncle Joe, we're here!" he called into the room. There was a sound of papers being shuffled in another room, a drawer shut, and presently a tall, broad-chested man entered the room.

"Ah, here you are!" he smiled. "So, you're Steve. I would've recognized you anywhere; you're an image of your father. How's your dad? I haven't seen him since the day we both graduated from college. Long time ago."

"He's fine, sir." Steve said, blushing. He was startled. He had not expected the Chief to look the way he did. In his mind, the Indian Chiefs were always attired in their tribal costumes, the way they were in the movies.

Chief Joseph, The Flying Raven, wore a dark gray business suit with a white shirt and striped tie, like the suits Steve's father owned. If anything, the Chief was dressed more formally, for Steve's father favored sneakers, jeans and T-shirts as his attire.

"Welcome!" the Chief said. "Tell me about yourself...your dad wrote that you were interested in Native American history. Is that why you want to spend your summer with us?"

"Yes, sir."

"Well, I'm glad. I'm sure your stay with us will answer many of your questions."

"You did not mention the *real* reason," Jim nudged Steve with his elbow. "Tell Uncle Joe about Sasquatch!"

Steve blushed. He swallowed hard and blurted out, "It has

been my life's ambition to find Sasquatch."

The Chief raised an eyebrow. "You, too, eh? It must be contagious, this quest for Sasquatch. My old college professor, Dr. Stone, spends several weeks with us every year, in search for Sasquatch. So far without any luck."

There was a noise on the porch. The dogs jumped and danced greeting a boy, crowding behind him in the doorway.

"Hi! I'm Joey, the Little Raven," he said stretching his hand.

Steve liked him at once. "Hi!" he said, "Glad to meet you, finally!"

"Me, too." Joey was about thirteen years old, Steve's own age. His shoulder-length hair was so black that it looked blue. It accentuated his bronze-colored face dominated by a pair of dark eyes, the color of black cherries. He wore cutoff jeans and a white T-shirt with a Harvard University logo. He was barefoot.

"Okay Joey, show Steve where he's to sleep, then wash up and come back for supper," the Chief said. "Are you staying, Jim?"

"No, thanks, Uncle Joe. I promised Sandy I'd take her to a monster movie, speaking of monsters!" he winked at Steve. "See you later!"

Joey led Steve to the back of the house to a small room with two bunk beds. The wide-opened window faced the broad meadow, beyond which stood the silent dark forest.

"Somewhere in the forest, Sasquatch is waiting," Steve thought.

If one were to ask Steve Bradley why he was so fascinated with the legend of Bigfoot, he wouldn't be able to answer. Ever since at the age of ten, when he had read for the first time about this half-man, half-ape creature, Steve decided that it was to be his destiny to discover it. He worried that someone may beat him to it before he was old enough to begin his quest. Periodically there were claims that someone had seen Sasquatch, although no one had ever presented proof of its existence. Anthropologists debated the validity of such claims, many even suggested a crude fraud, but to Steve, Sasquatch was real. He was there, somewhere in the primeval forests of the Northwest, waiting for Steve to discover him.

Steve had read everything he could find on the subject of Bigfoot, cutting out articles and pictures that appeared in magazines and newspapers, compiling a thick file. Artists' sketches always depicted Sasquatch as a hairy monster, huge and menacing, with long apelike arms, a small head and a broad chest and shoulders. His round, small eyes, all but hidden under his bulging brow with its tufts of coarse hair, would peer from these drawings, as if challenging Steve.

The creature was described as weighing from six to eight hundred pounds and its height was approximated at six to eight feet.

"Do you suppose, Sasquatch is there?" Steve said pointing to the woods.

"Who knows?" Joey shrugged. "My dad thinks it's a lot of baloney, but the old folks believe in Sasquatch with all their hearts."

"What about you? Do *you* believe that he's there?"

"I don't know. It's hard to believe in something that your own father makes fun of..."

"Yeah, I know. My dad makes fun of Sasquatch also. And of me, for believing that Sasquatch exists. But wouldn't it be great to prove to everyone how wrong they are?"

"It sure would. Anyway, let's go and eat. I'm starving!"

The Chief and the boys settled around the table in the kitchen as an old woman slowly stirred the fish stew in a large kettle. Just as slowly she ladled the steaming food into three deep bowls and placed them on the table before the Chief and the boys.

She was dressed in a dark, voluminous, calico skirt and a red blouse. She wore her gray hair plaited into two thin braids which framed her flat, wrinkled face. The braids fell to her chest, intermingling with necklaces of beads and seashells.

The Chief removed his tie and jacket and rolled up his sleeves. His hulking frame seemed to fill the whole kitchen. Steve could see his biceps play under his rolled-up sleeves.

"Okay, fellows, dig in," he said. "Steve must be hungry after his long trip. And you, Joey, I know you. You're *always* hungry!"

Joey grinned. "I'm a growing boy!"

The Chief continued, "You're probably wondering, Steve, why there are no traces of a woman living in this house. I am a widower. My mother keeps house for Joey and me but she prefers to live by herself in the same place where she lived when my father was alive. Anyway, to change the subject, Dr. Stone is arriving in a couple of days to resume his annual search for Bigfoot. Down your alley, eh Steve?" he smiled. "Dr. Stone will stay with us, as usual," he continued, growing serious.

"You'll like him," Joey said. "He's a regular guy, he's lots of fun. He likes to tell stories about how things were thousands of years ago. He's an anthro...something..."

"An anthropologist," his father helped him. "He's one of the best authorities in the country on old civilizations."

"But he's not at all stuffy," Joey explained, afraid that Steve might find Dr. Stone's expertise too dull. "He knows *everything* about monsters and devils and human sacrifices," he boasted.

"Yes, I've heard about him. My father has several books by Dr. Stone in his library."

"I bet he's going to write a book about Sasquatch, too!" Joey said handing his bowl to his grandmother for another helping of the stew. "If he can find one!"

"No need to find..." the grandmother said suddenly. "Sasquatch is there," she pointed to the forest with her wooden ladle. "I saw him yesterday."

"You what?" Steve jumped up, upsetting his chair. "You saw Sasquatch yesterday?"

"I sure did. I saw him as clear as I see you now."

Steve felt blood rush to his face and beat violently against

his temples.

"Where did you see him Mother?" the Chief said gently.

"Right there, in the woods. I was gathering berries and he was feeding on the berries not far away."

"What did you do?" Steve said almost in a whisper, his voice suddenly failing him.

"Nothing. He don't bother me none, and I don't bother him none. We go our own ways..." She turned back to the stove, rearranging the pots in her unhurried way.

"Well, there you are," the Chief turned to Steve. "However, the question still remains. What *exactly* did my mother see? Was it a bear on his hind legs, eating berries?"

"It was Sasquatch," the old woman said without turning from the stove.

II

New Friends

Steve lay on the upper bunk, wide-awake, the sleep evading him. The moonlight flooded the room, making all objects stand out sharply against the shadows. The intoxicating aroma of wild flowers mixed with the salty breeze from the ocean and wafted in through the open window.

Steve stretched and turned his pillow over. He enjoyed the nocturnal solitude. The impressions of his first day on the reservation crowded his mind, keeping him hyped up beyond the practical necessity for sleep.

"Boy, what a cool guy Joey's father turned out to be! A lawyer and a Chief! I wish my dad were here and could see him now!" Steve thought.

Steve's father and the Chief used to be classmates at UCLA. They kept in touch with one another from time to time, but had not seen each other since their graduation, their various careers taking one of them to the Harvard Law School and the other to Hollywood.

Steve Bradley, Sr. was a screenwriter. He worked in his office above the garage, joining his family only for meals. When he had to meet a deadline, he often slept in his office as well, emerging from it two or three days later unshaven and irritable. On such days Steve and his mother stayed out of his way.

Steve often thought that he would rather his father worked at a gas station and kept regular hours. He missed his father's company. He cherished their rare moments together when his father was free of tension to engage in horseplay, like racing with him in the pool. He treasured the infrequent times spent in his father's book-lined office, talking, laughing and making up silly jokes.

But one subject became unspoken between them—Sasquatch. Ever since Steve had mentioned his fascination with Bigfoot, his father barraged Steve with stinging ridicule. "Bigfoot! Loch Ness Monster! The Abominable Snowman of the Himalayas! Only idiots believe in such nonsense!"

"But Dad, there are scientific expeditions organized by several countries in search of these creatures!"

"A bunch of fools squandering taxpayers money!"

"But Dad, don't you think that it is *possible* that these beasts might exist?" Steve had insisted.

"No," his father replied categorically. "As for you, spend your time on something more useful, like getting better grades in algebra, instead of filling your head with fantasies."

Steve blushed painfully. It was true, his grades in algebra dipped down to a C, still, his father should have been more sensitive, Steve thought. His interest in Bigfoot had nothing to do with his grades in algebra! After that day, his father teased him mercilessly about the "monsters." Steve began to avoid his father.

But he continued to dream about Sasquatch. Then, one morning in spring, Steve's father came into his room with a newspaper clipping.

"There are some interesting excavations going on," he began. "They have uncovered a whole Indian village. I thought it might be interesting for you to witness the work of archaeologists." He paused, looking at Steve owlishly through his horn-rimmed glasses. "Besides," he continued matter-of-factly, "the digs are being done around the area where your friend Bigfoot supposedly has been seen again," he grinned. "So, how about it, tiger?"

"Gee, Dad, that would be great!" Steve jumped to his feet. "Okay, I'll get in touch with my old college roommate. He's a Chief of a small tribe over there. Perhaps, we can arrange for you to spend a few weeks at his house."

"Gee, thanks, I'd like it very much!"

"I thought you'd like the idea," his father smiled and ruffled his hair.

Steve easily adapted to the routine of the Chief's household. In the mornings, he and Joey were up early to have breakfast with the Chief, who would leave promptly at seven-thirty for his office in Olympia, the state capital. He kept another office in his house on the reservation.

After the Chief was gone, the boys were on their own. Steve wished they would be allowed to go to the digs at Lake Ozette, but the security was very strict and no minors were permitted on the territory of the excavations. Disappointed though he was, Steve had plenty to do just following Joey around. They fished from a small pier, they swam in the ocean and surfed on the homemade surfboards, which skimmed the waves better than any boards Steve had used before on the beaches of California.

However, he especially liked visiting Joey's private island. At one time this island had been a part of a high cliff, but the pounding ocean, heavy rains and winds separated it from the mainland along with dozens of similar craggy formations. The coast along Cape Flattery was dotted with these peculiar vertical islands protruding perpendicularly, pointing to the sky like index fingers, coniferous trees growing out of every crack of their rocky surfaces. During low tide it was possible to wade out to many of them, but when the tide was high, the waves churned and crushed at their foundations.

One of these islands Joey claimed as his own. It was no bigger at the top than his own room back in the village, but it rose above the ocean for almost a hundred feet.

Access to the summit was extremely dangerous. The menacing rocks at the foot of the island waited for just one wrong step to receive a fallen body, to tear it to shreds in the churning ocean as it surged against the rocks, breaking its waves into myriads of sparkling sprays.

Joey perfected scaling the rock by attaching a stout rope to a tree trunk on the summit. He demonstrated to Steve how to use the guide rope while probing with his bare feet for every indentation in the craggy surface.

Steve followed Joey up the vertical wall of the rock, trying to block out of his mind the angry surf and the sharp rocks waiting below.

"Keep going, don't look down, you're doing fine!" Joey shouted from the summit. When at last Steve's head was level with the summit, Joey stretched his hand out, pulling him up. Steve collapsed on the mossy ground, breathing heavily.

"Were you scared?" Joey asked.

"Out of my wits!"

Joey laughed. "Me, too. But it's worth it, no?"

Steve looked around. The dark green ocean was at his feet. He could see the kelp beds as they swayed under the water back and forth, like a forest in the wind. Far on the horizon a gleaming white cruise ship was moving north toward Alaska. The only sounds he could hear were the sharp cries of the sea gulls and the incessant rumble of the surf at the foot of Joey's island.

"Neat place, eh?" Joe grinned. "I'M THE KING OF THE MOUNTAIN!" he suddenly yelled, the echo faintly acknowledging his claim.

The first week of Steve's vacation flew faster than any time in his entire life. He and Joey were inseparable. They looked so different: Steve, blue-eyed, too tall for his age, with gangly arms and legs and large feet, with curly, blond hair, bleached almost white by the strong California sun, and Joey, stocky, with black eyes and thick eyebrows grown together on the bridge of his finely shaped aquiline nose. Steve's fairness of complexion complemented Joey's swarthy good looks.

Joey enjoyed showing his new pal off to the other boys on the reservation. "His dad writes for TV!" he bragged. "And his mom knows all the movie stars!"

Steve never thought there was anything significant in his father's writing for television. It was his profession. Neither was there anything special in knowing movie stars. Most of his parents' friends were in "the industry," writers, directors, actors; and Steve grew up seeing their famous faces in his house. He was even named after one of them—Steve Allen.

But the boys on the reservation venerated television and movie heroes. They showered Steve with questions about Sylvester Stallone, Clint Eastwood and Arnold Schwarzenegger, selecting the toughest characters among the movies stars. They wanted to know what kind of cars they drove, what kind of money they made and whether they were really "tough." It amused Steve that his new friends were so taken with their bigger-than-life heroes that they envied *him* for knowing them! These kids, who had all kinds of adventures right on their doorstep envied *him*—a city kid!

But not all the boys, however, accepted Steve. Billy Wallace, nicknamed Chickie, one of Joey's numerous cousins, refused to welcome Steve.

"Stay out of his way," Joey advised. "He's mean. I can't stand him."

"Why?"

"Because he's a cheat and a thief. I hate him, even though he's my cousin. It all started a couple of years ago. Billy Wallace and I are in the same class. One morning, on the bus, he asked to look at my homework. We had an assignment to write a composition. So, I let him read it. A few days later the teacher called us both in. She said that our compositions were practically the same."

"He copied your composition!"

"He changed a few things, but he had used my story."

"Did you tell the teacher that he had read your composition?"

"Of course not! It was *up to him* to admit that he had used my story. But he said nothing. We both received a D as a punishment, but the teacher said that one composition was well-written and deserved at least a B+."

"Yours."

"Yes. The teacher knew that it was mine, but she wanted us to confess. Chickie said nothing. He dishonored himself. All my friends know about it. They began to call him Lying Chicken instead of his native name, Flying Hawk. He hates his nickname! But it stuck to him. The kids shortened it to 'Chickie.' Now, even the grownups call him by this name. He's a mean dude. Don't mess with him."

Steve tried to stay out of Chickie's way, but Chickie took every opportunity to taunt him by calling him White Belly.

Steve had a premonition that he and Chickie were heading for a showdown. He dreaded that possibility.

The Chief arrived from his office in Olympia and allowed Joey to park his car in the garage.

"Herb Stone should be here in a couple of days. Run to Jim's trailer and let him know. He's supposed to prepare the community center basement for Dr. Stone's equipment."

The boys rushed toward the cliff and its sharply descending path to the beach. Down below, among the driftwood and kelp brought ashore by the tides, stood several trailers without wheels, propped on the cement blocks just above the markings of the high tides. Between the trailers there were faintly smoldering blackened piles of rocks where the occupants kept their cooking fires. Many elderly Indians who lived in the trailers preferred to cook over an open fire, like in the old days.

Jim flung the door of his trailer. "What are you guys doing here?"

"Dad sent us. He wants you to fix the community center basement for Dr. Stone's stuff. He's arriving in a couple of days," Joey said.

"It's all done. Sandy and I cleaned the whole messy place last night and I even unplugged the toilets!"

"Stop bragging!" Sandy, Jim's girlfriend, came to the door.

"Meet my friend Steve Bradley," Joey said.

"Glad to meet you," she smiled, exposing a row of even white teeth. Steve found himself blushing painfully under his suntan.

"Hi!" He shook her small, strong hand. He felt awkward, not knowing what else to say. He always felt shy with girls, his shyness increasing in proportion to a girl's attractiveness. With Sandy he was speechless.

Sandra Williams, Moonglow Daughter, was nineteen and in the full bloom of her beauty. She was quite tall and slender, with bronze-colored skin which Steve's mother tried to achieve every summer with the help of expensive creams and lotions.

Sandy had high cheekbones and her thin-nostrilled nose looked delicate, as if chiseled. Her full lips reminded Steve of ripe red plums. She wore bright beads plaited into her shiny black hair which fell on her shoulders in two thick braids.

"She looks like an Indian princess," Steve thought. He realized that he probably looked foolish, unable to think of anything to say. He blushed even deeper.

"Come in, come in!" Sandy smiled. She was accustomed to people's first reaction to her striking beauty. "Wanna Coke?"

"Sure," Joey said.

"What about you?" Sandy turned to Steve.

"Me, too," he said, his voice cracking, his face red.

Sandy went to the small ice chest and took out two cans. "The ice is all melted," she complained. "Sorry, the Cokes aren't cold enough."

"I don't mind. It's okay," Steve said, clearing his throat.

"You see, Jim has no electricity in his trailer," Sandy explained. "While it might be romantic in the movies, candlelight and all, it's not very practical. I bring him a bucket of ice each time I come visiting. In the summer it's easy. I just empty my folks' fridge, but in the winter, when I'm not here..."

"In the winter it's cold enough. I don't need ice in the winter," Jim said opening another can for himself.

"Where are you in the winter?" Steve dared to ask.

"I go to school in California. I live on the campus."

"Sandy is going to be a teacher." Jim said.

"I wish you were *my* teacher," Steve blurted out, and blushed crimson again, embarrassed.

"Me too," Joey joined him.

"Me too," Jim said. Sandy laughed. "Thanks, boys, but by the time I start teaching high school, you'll be too old.... All of you, too old," she said pretending not to notice Steve's red face. "You'll be in college, studying to become a...what? What do you want to be, Steve?"

"He wants to be a discoverer of Sasquatch!" Jim said.

"Yeah? Around here everyone wants to be the discoverer of Sasquatch."

"You too?" Steve stammered.

"Sure, me, too."

Steve laughed happily, suddenly at ease with her. Somehow, Sandy made him feel "one of the crowd."

III

Where's the Proof, Dr. Stone?

Dr. Herbert Stone and his students arrived three days later. Steve watched as the two jeeps proceeded slowly along the main street toward the community center. Dr. Stone was seated next to the driver in the leading car. He was a thin man of about sixty, with a neatly trimmed gray beard and mustache, wearing a pith helmet and an old, beige sweater with a large hole at the elbow. The driver of the car was a young man with long, unkempt, curly hair and struggling face whiskers.

The driver of the other jeep was a redheaded girl. On the passenger seat next to her rode a magnificent St. Bernard dog, looming majestically over his diminutive driver. On the back seat, dwarfed by the huge dog, rode a young black man, his chest criss-crossed with leather straps of two cameras, reminding Steve of a "bandido" of the old Western movies. The bandidos wore their bandoleers of bullets in a similar fashion.

A group of children surrounded the jeeps, gaping at the new arrivals.

"Hi, Dr. Stone!" Joey yelled as he ran to the leading car.

"Hello, Joey, my boy! How are you? How you've grown! I would've never recognized you!" Dr. Stone smiled as he climbed out of the car.

"Hey, Steve, where are you?" Joey called. Steve stepped

out from behind the crowd of children.

"How do you do, sir." He shook Dr. Stone's bony hand.

The members of the expedition stepped out of the cars, stretching their limbs. The red-headed young woman held tightly to the dog's collar but he had no intention of leaving her. He collapsed on the ground, like a sack of flour, and let out a deep sigh of contentment, making the children giggle. They had never seen such a large dog.

"Okay, Joey, where do we go now? We're at your mercy," Dr. Stone said.

"You stay at our house. The others—with the folks in the village. Here's Jim. He's in charge. He'll show your friends where to go." Joey was glad that Jim was there to take over the arrangements.

"Hello, Dr. Stone! We're sure happy that you're back!" Sandy greeted the professor.

"Thank you, my dear. Glad to be back. You're more beautiful than ever! May I patent the secrets of your beauty for mass distribution and for my ultimate rise to fame and fortune on the strength of it?" The young people in Dr. Stone's entourage broke into laughter joined by Sandy and Jim.

"He's a character," Joey whispered to Steve. "He talks funny!"

"How are you, Chief?" Dr. Stone turned to Jim. Then, with a solemn expression on his bearded face he announced to his group, "This is Jim Brown, Raven's Wing. He'll be the next Chief of this clan. Treat him with the *greatest* reverence!" Everyone laughed again, as Sandy and Jim shook hands with the new arrivals, Lucy Warner, Dave Rosen and Mike Powers.

"Okay, Joey, lead the way! I must pay my respects to your grandmother or I'll be in the doghouse with our Tiny here!" He pointed to the dog who perked his ears up hearing his name, but relaxed again when no command followed.

The boys, each carrying one of Dr. Stone's duffel bags, led their guest to the Chief's house.

"Well, what's new, Little Raven?" Dr. Stone said.

"Nothing much...except that about ten days ago Grandma saw Sasquatch." As Joey had anticipated, the news had a galvanizing effect on Dr. Stone. He stopped, causing the crowd of children that followed closely at his heels to collide with him.

"What? She saw Sasquatch? Where?"

"In our back yard. In the meadow, at the edge of the woods," Joey said with a deliberate shrug, throwing Steve a mischievous glance.

"Has anyone followed him?" Dr. Stone's face was very serious.

"No. We waited for you. Dad said that no one was to go to the woods until you get here."

"Very considerate of him. We don't want to scare Bigfoot off!"

They reached the Chief's house. The dogs rushed at Dr. Stone, barking furiously, but Joey ordered them to lie down. They obeyed instantly, and, Steve thought, with relief. They hated to be disturbed.

"Dad isn't home yet," Joey said holding the door for Dr. Stone to enter. "I'll tell Grandma that you're here," he disappeared in the passage leading to the back of the house.

Dr. Stone looked around. The front parlor felt cool after the hot afternoon sun. He plunged himself into an armchair. "Ah...this feels good!"

Steve sat on the chair across from the professor trying to think of something clever to say. He felt awkward, as he usually felt with people whom he especially wanted to impress. He stared at his huge bare feet, embarrassed that they were dirty. He tried to hide them under the chair.

"Well, Steve, how do you like what you found here so far? It's not what you expected the Indian village would look like, eh? Did you ever think that Indians would live just like you and me, in ordinary houses with TV and washing machines rather than in teepees, or, in the case of the Northwest, in kwan houses, eh?"

"No, sir. I always thought of them as I saw them in the movies," Steve said blushing.

"Are you disappointed?"

"Oh, no, sir, I like that the Native Americans are no different from the rest of us." Steve wanted to elaborate on his statement, but no right words came to his mind. He felt stupid.

"Here is Grandma!" Joey announced from the door, rescuing Steve. Dr. Stone jumped to his feet to greet the old woman as she entered the room.

"Cooing Dove!" He embraced her and all but lifted her off her feet. "You look prettier every year!"

Her face broke out in a smile of pleasure exposing several gaps in her mouth where her teeth were missing.

"Cooing Dove!" Steve suppressed an impulse to laugh. "What a name for an old crone!" he thought. But at once he felt ashamed. It's not her fault that she grew old with a young name...she's not an 'old crone'...she's a cool old lady and I like her.

"Let me look at you!" continued Dr. Stone holding her at arm's length by her shoulders and peering earnestly into her face. "You look great! What do I hear? You saw Sasquatch?"

Grandma nodded. "Yeah. Ten moons while."

"Tell me about it." Dr. Stone was barely able to contain his excitement. He trembled, and to cover up, he coughed.

"Nothing to tell. I pick berries and he's there. Eating berries. He look at me—I look at him. He grunt and move away. I pick berries and go home."

"Are you *positively sure,* beyond a reasonable doubt, Grandma, it was not a bear?" Joey sounded like Perry Mason, Steve thought.

"I no fool. I know bear. It was Sasquatch," she replied with dignity. "I go cook your supper."

"Stay, stay!" Dr. Stone implored, but grandma was offended. She left the room.

Dr. Stone made a gesture of despair with his hands. "You

see what you've done? We have to wait now until she calms down and is willing to show us where her encounter had occurred," he said.

"I'm sorry, Uncle Herbie," Joey looked crestfallen. "I just asked her the same question as my dad did..."

"Never mind. What she would accept from your dad, she would not tolerate from a youngster. You, Little Raven, of all people, should know that," he said not hiding his annoyance.

"I'm sorry," Joey mumbled, blushing deeply. Steve wished he could come to his help, but nothing would come to his mind.

"Tell it to your grandma," Dr. Stone was unmollified.

There was a sound of a car driving into a garage. The Chief had arrived home.

It was late in the evening after the supper when Dr. Stone and the Chief settled themselves on two creaking rockers on the back porch, reminiscing. They recalled the names of their mutual friends and colleagues, asking one another what ever happened to so and so, the unknown fascinating people becoming real to Steve.

Finally, Dr. Stone stood up and stretched his arms above his head until the joints cracked. He yawned loudly with obvious relish.

"Time to go beddy-bye..." he said. "I was up at five this morning and I feel a bit tired. Tomorrow I will have a talk with Cooing Dove and perhaps entice her to show me where she saw Sasquatch. Do you think she will do it?"

The Chief shrugged. "For you—she'll do it. For me, never. She's angry with me. It is supposed to be very good omen for the tribe to have Sasquatch nearby, but instead of rejoicing at her good news, I was skeptical of her story. You see, my mother, bless her heart, has ambiguous feelings about me. On the one hand, she is proud that I am successful in a white man's world, while on the other, she scorns me for not following the old native ways."

Dr. Stone sat down again, his tiredness gone. "I know exactly what you mean," he said. "But why do you refuse to believe in Sasquatch? Most of your people, young and old, are convinced that Bigfoot lives right there, in those woods. Why not you?"

The Chief laughed. "That's what I call turning the tables on me! Here you are, a learned professor, who is supposed to be skeptical about unproven theories. Yet, you of all people, go chasing after some mythological creature, asking me to agree with my mother, an illiterate woman, that such a creature exists on my very doorstep! All you have as *proof* of Sasquatch's existence are stories told by some old natives.... Why are there no bones of Sasquatch? They die, so there ought to be some bones left somewhere...or fossils, or some other archeological evidence..."

"Oh, shut up!" Dr. Stone laughed. "It's true, we can't find any bones, but you know as well as I that there are many eyewitness reports about seeing Sasquatch. Aside from your mother's!"

"But no bones!"

Joey nudged Steve with his elbow. "The fun is just beginning!" he whispered. "I just *love* their arguments!"

It was getting dark. Their faces began to lose their features, gradually becoming vague masks. Soon only the Chief's white shirt indicated where he was sitting. The rest of them, dressed in darker clothing, were swallowed by the gathering night.

There was no moon. The air felt oppressive, as if a storm was brewing. No breeze came from the ocean, no whisper of wind among the trees. Steve thought how very appropriate the heavy atmosphere was for the discussion about the mysteries of the rain forest.

"Do you want me to recount how many sightings of Sasquatch there have been in the last decade?" Dr. Stone asked.

"No, no, please, spare us the tedious statistics!" the Chief cried out in feigned horror.

"Please, Uncle Herbie, tell us! Don't listen to Dad, tell us!" Joey pleaded.

"Yes, Dr. Stone, please, tell us," Steve joined.

"Okay...just for Joey and Steve. You, Big Joe, plug your ears. To start with, do you know when the gorilla was first discovered?"

"*Gorilla?* What has gorilla got to do with the sightings of Sasquatch?" The Chief was sarcastic.

"Everything. And nothing. What I mean is that the discovery of gorillas has direct relation to our *attitudes* about Sasquatch and other strange phenomena," Dr. Stone continued. "For centuries the natives of Africa knew about gorillas. They saw them. They described them to the so-called civilized men, but no one believed the natives. The white men laughed at them. They ridiculed their stories very much as we do it today with Sasquatch. The white men laughed at the natives who insisted that there were giant, manlike beasts, who walked upright, who had hands and fingers like men and who nursed their babes like human mothers. Very much indeed as we laugh at those who say that they saw Sasquatch. Like your own mother."

"Well, I don't ridicule anyone. I just don't believe that my mother, or anyone else for that matter, saw *Sasquatch*. More likely she—and the others—saw a large bear, standing on its hind legs." the Chief said.

"Tell us more about the gorillas," Joey begged. "Who was it, who finally believed the Natives?"

"It was Paul Du Chaillu, a French-American explorer. He had spent decades exploring Africa, and among other things, he was the first white man to come upon a gorilla. He called the beast 'the hellish dream creature.' The Natives caught it for him and Du Chaillu brought the gorilla to America for exhibition. Do you realize that it was less than one hundred and fifty years ago? Gorillas roamed Africa for centuries, yet their existence became known to the rest of the world only in 1856? Doesn't it make one think that it is possible, just *possible*, that somewhere in the mighty wilderness of California and Oregon mountains, or in the rain forests of Washington and wild canyons of British Columbia, there are creatures, still undiscovered? After all, there are over a *hundred and twenty-five thousand square miles* of unexplored wilderness! Couldn't it be that Bigfoot is there, somewhere inside this unexplored area, still unseen by anyone but a few..."

Steve heard the Chief strike a match. For an instant his face was illuminated by the flame of the match, as he lighted a hurricane lamp on the porch table.

"Or, take another example," continued Dr. Stone. "Our giant redwoods, the Sequoias. They are thousands of years old, yet, they were seen by the white man for the first time less than two hundred years ago! There are still a lot of things that we had never seen in this world. Lots of things to search for and to discover!"

Steve felt the shiver of excitement run down his spine. "Yes," he thought. "There are things for me to search for and to discover...I know *exactly* what Dr. Stone means!"

"This is why I want to spend the next several weeks in Eagle Canyon," Dr. Stone continued as if reading Steve's mind. "My young colleagues and I will be looking for footprints and if we're lucky, we might come upon the real thing, Mr. Sasquatch himself!"

"Well, I sure wish you luck!" the Chief said. "You'll need it. The time is ripe for someone, *anyone*, to come up with indisputable proof. Or—shut up." He paused and poured himself a glass of water. "Of course, I don't mean it personally."

"I saw a lot of pictures of Sasquatch's footprints." Joey said. "And there was a show on TV about him. But it looked phony," he added.

"I examined the casts of the footprints," Dr. Stone said. "I also interviewed dozens of people who claimed that they saw the beasts. One of the most believable accounts however was a film made in 1967 by Roger Patterson at Bluff Creek in Northern California."

"I saw that film on television!" Steve cried, unable to control his excitement. "Bigfoot looked like a huge, hairy man! And you should have seen his footprints! They were enormous!"

"A man in a gorilla suit, more likely. Someone's pulling your leg, gentlemen. Anyone could have easily forged the footprints!" the Chief laughed.

"I am convinced that it would be *impossible* to forge all the existing footprints," Dr. Stone said. "Granted, some of them could have been made by pranksters, but I insist it's *impossible* to forge them all because they are found in so many places, long distances from one another...also, the forgers would have to be very well acquainted with the anatomy of a foot, with its orthopedics. I can't conceive of some mad orthopedist rushing through thousands of miles of mountain wilderness, stamping the ground with giant footprints!"

"Maybe he flies in a helicopter," Joey suggested.

"Now, really, Joey, you can't be serious! Some of the footprints were found years ago, long before the invention of helicopters. Have any of you ever read Teddy Roosevelt's book, *The Wilderness Hunter*? It was published in 1893, long before any of us were born, and I might add, long before the invention of helicopters. Read it! You'll find it very interesting. Our famous president described Sasquatch in detail."

"Had he actually seen him?" Chief asked with interest.

"No, I must sadly admit. Teddy Roosevelt wrote that he had heard about Sasquatch from some hunters and gold prospectors who had seen him."

"Aha! Again, *he had not seen Sasquatch!* It's always someone else who had supposedly seen him. Never the person himself!" the Chief exclaimed in triumph.

"Yes, unfortunately it is so," Dr. Stone agreed. "That's why I have spent my summers in the rain forest for the past ten years, trying to find a definite proof of Sasquatch's existence. Because I believe in him with my whole heart!"

"You're an incorrigible romantic. All I can say is good luck!" the Chief said getting up. "It's late. I'm in court tomorrow."

"And I must unpack my gear and set up the laboratory. This conversation only whetted my appetite. Perhaps *this* is the year when I'll find Sasquatch!"

"And high time, too!" the Chief said amicably. "Good night."

IV

Visit to the Nootkas

The boys left the house early next morning, hurrying after Dr. Stone to the community center basement.

At once fog enveloped them in its misty embrace, obscuring their vision. The grass felt cold and wet under Steve's bare feet and he was tempted to return for his sneakers. But Joey seemed to be unaffected by the chilly grass. Steve followed him, shivering, unwilling to appear less rugged than his friend.

"Ah, here you are!" Dr. Stone greeted them. "Help Lucy and the guys unpack, will you?" He pointed to a pile of cartons.

"Hi!" Lucy smiled at the boys. She looked very young, barely older than themselves. She was pretty, her round face covered with freckles and her coppery hair, cut very short, giving her a boyish look. She was wearing baggy gym pants and a yellow T-shirt with a giant footprint of Sasquatch emblazoned across the chest.

"Lucy will stay here at our headquarters while Dave, Mike and I go into the bush with our cameras," Dr. Stone said. "If we're lucky, we'll bring back the films for Lucy to develop. If you boys know anything about photography, I would appreciate your giving Lucy a helping hand. Become her assistants."

"Gee, thanks, Uncle Herbie. We'd love to!" Joey said. Steve thought that he would rather follow Dr. Stone into the forest, but he felt too shy to ask whether he could.

"Don't even think it," Joey whispered, reading his mind. "I tried it last year and the year before, and the answer was always no! Working in the darkroom could be fun."

"I suppose.... At least we'll be the first to see Dr. Stone's photographs. Besides, I'll be with Lucy," Steve thought, his spirits rising. He knew that he was falling in love with Lucy, as he did with every pretty girl that he met, regardless of age.

The boys began opening cartons containing the walkie-talkies and a small radio transmitter.

"Where's the *odorometer*?" Dr. Stone suddenly thundered, making everyone jump. "How am I supposed to measure Sasquatch's stink without the odorometer?"

"It must be in one of those cartons. I packed it myself." Lucy replied calmly.

"It had better be. Or off with your head!" the professor growled ferociously. Everyone laughed.

The expedition was well equipped with a portable video and regular cameras with telescopic and wide-angle lenses designed to be used in the dark.

Steve was impressed with the amount of equipment needed for the expedition. Dozens of rolls of film, light aluminum frame tents, inflatable mattresses, goose down sleeping bags, Coleman stove and lanterns, water purification tablets and rolls of toilet paper. There were picks and shovels and soft whisk brooms to dust off the dirt should the explorers come upon some fossils. There was even a microscope and a box of glass slides.

"Why do you need a microscope, Uncle Herbie?" Joey asked. "If you find Sasquatch you won't need a microscope to see him. He's *big*!"

Dr. Stone scowled. "Fun-ny! That's what I call a good joke! Ha-ha! Don't you realize that we might find proof of Sasquatch's presence even without seeing him? Things like tufts of hair stuck to the branches, or marks of teeth on tree bark, or perhaps even some droppings. How would we know that the hair did not belong to a bear? Or whether the droppings were from animal or human species without examining them under a microscope?"

"Boy, he thought of everything!" Steve thought.

"But what will you do if you find nothing?" Joey insisted.

"I'll come back next year. And the next," Dr. Stone replied.

The expedition was ready for departure. The boys helped Dr. Stone and his assistants to strap the heavy packs to their shoulders. The village children ran after the explorers to the edge of the meadow but Joey and Steve did not join the children. They were now Dr. Stone's assistants and their place was in the lab, as the community center basement became known thenceforth. They returned to the lab.

Next morning, the village bubbled with excitement. The canoes resting in front of the houses were turned over and the tribal designs on their hulls touched up with fresh paints. The women scurried in and out of the houses with armloads of festive costumes.

In the Chief's house, Grandma rummaged in large bentwood boxes decorated with ravens and whales, the symbols of her clan, taking out ceremonial garments embroidered with glass beads and rows of pearly buttons.

"What's going on?" Steve asked.

"We're going to a potlatch!"

"Potlatch! Gee, I always wanted to see a potlatch!" Steve exclaimed.

Grandma brought in the Chief's ceremonial dress. Steve watched Joey's father try it on. Suddenly the urbane lawyer was gone. In front of the mirror there stood the *Chief of the Ravens*, awe-inspiring even though the gray, flannel slacks were still visible under his native robes.

"I am looking forward to this visit to the Nootka tribe," the Chief said adjusting a cedar bark mantle over his broad shoulders. "Nothing like a good potlatch to bring old friends together. I wish though that you boys could have seen the *real* ceremonial visit," he added with a nostalgic smile. "Hundreds of canoes would be gathered, traveling sometimes for great distances to visit a friendly tribe. Nowadays we go in our power boats. Sometimes we even charter a steamer with regular cabins and toilets!"

"But I saw the canoes being readied," Steve said.

"Oh, yes, we'll take them along. A couple of miles before landing we'll anchor our fishing boats and pile up in the canoes. We'll arrive in the old style, but it will be a *reenactment* of our past. And so will be the potlatch itself. It will be a reenactment. It won't be exactly as it once was."

"Never mind how it once was," Joey said. "It's still lots of fun. Steve'll love it! And in any case, I prefer to go by a power boat. It's much faster and I don't have to paddle for hours!"

The Chief signed with mock resignation. "Oh, youth, youth. You never appreciate the past, until it's too late. Soon there will be nothing left of our heritage but artifacts in museums."

"I like the past," Grandma said from the kitchen door. "I like better when our clothes were made of cedar bark like in the old times. They last longer," she said. Grandma was attired in a long shift made of stiff cedar fibers. Her shoulders were draped in a large triangular shawl decorated with designs of a raven, a whale and a frog, outlined with hundreds of shiny mother-of-pearl buttons. Steve recognized her shawl as a Chilkat blanket. He had seen a similar one in a museum.

Grandma wore a headband woven of leather strips and bright beads. Heavy seashell pendants dangled on her ear lobes. But the most striking part of her appearance was the *labret*, a wooden plug, the size of a silver dollar, stuck inside her lower lip. It deformed her face, exposing her few surviving yellowed teeth. The old woman carried herself with great dignity, proud of her appearance.

"Mother of mine, you truly look like the mother of the Chief ought to look! You make us proud of you!" the Chief smiled at her.

Steve glanced covertly at Joey, expecting him to snicker at the *labret,* but Joey regarded his grandmother with the same expression of pride.

"You look great, Grandma," he said.

The old woman nodded regally as she returned to the kitchen.

The villagers were ready to depart on their journey to the Nootka tribe in British Columbia.

"Oh, how I wish I could go with you!" Lucy sighed. "But, I must wait for Dr. Stone's messages..."

"We'll bring you something from the potlatch," Joey said.

"And we'll take snapshots," Steve promised. They waved good-bye and ran toward the boat landing.

Dozens of fishing boats idled at the pier in the tiny harbor. The canoes, looking like shiny, black fish scattered about by a tidal wave, were resting on their decks.

The villagers boarded the boats, crowding down into the tiny cabins, packing tightly along the decks, waiting for the Chief's signal to depart.

Joey's father, wearing an embroidered long shirt and trousers of soft animal skins, stood alone at the end of the pier waiting for his people to be settled. At a respectful distance from him stood Jim and several other men highly placed in the tribe. They were to sail on the Chief's boat, as were Joey and Steve.

Chickie was herded along with women and children into one of the other boats. His face bore a dark, angry look. Steve sensed that he was churning with envy that Steve, a stranger, was invited to ride with the Chief.

"That kid sure sends some bad vibes," Steve whispered to Joey.

"You're not kidding."

One man stood out of the crowd. Steve recognized him as an old man who lived in a ramshackle lean-to on the outskirts of the village. Joey had told him that he was a witch doctor, the shaman.

The shaman began to sing in a high-pitched, nasal voice, stomping his bare feet on a splintered boardwalk while beating a drum. The seashell bangles around his ankles clicked following his every move. His long, gray mane, reaching almost down to his waist, was sprinkled with ashes and covered with goose down. It was matted as if he had never combed or washed his hair. He wore a headdress of tree bark and feathers, and around his ropy neck dangled several necklaces made from the beaks of puffins. The shaman's torso was bare. Every rib and bone of his chest protruded sharply and he had the wiry look of a disjointed puppet. He wore only a short loincloth, and his buttocks were exposed.

"He looks exactly as the shamans used to look in the old days," Joey whispered. "They used to believe that their magical powers were in their hair. That's why shamans never cut their hair. Nor comb it. Also, they never wash."

"I bet they stink to high heaven," Steve grinned.

"They sure do," Joey giggled, glancing furtively in the shaman's direction.

"But what do shamans *do*?"

"They perform ceremonial duties. People say that they communicate with the supernatural...they can stop bleeding. But dad says that the shamans' days are numbered. The young people don't want to be witch doctors. They would rather be *real* doctors."

"He's cool!" Steve said. "I like him."

"Don't look at him. Don't *ever* look directly into a shaman's eyes," Joey whispered. "He can put a curse on you!"

"You're kidding, aren't you?"

"No I'm not. A shaman's magic is powerful. Just don't look into his eyes."

"Okay. But of course I don't believe in magic," Steve added, sounding not too convincing.

"Never mind. Just don't look into his eyes!"

The shaman shook his rattles violently and the captains revved their engines. The Chief stepped down into the leading boat and his retinue followed him. The shaman leapt off the pier with the agility of a young man, landing in the Chief's boat. He crouched at the Chief's feet, and resumed beating his drum.

Steve craned his neck, fascinated by the shaman.

"He looks ferocious," he thought. The shaman's face was painted in black stripes with touches of red, beneath which Steve could see traces of an old tattoo. But the oddest of all—the shaman was wearing sunglasses!

"Bizarre! I like the old guy. I don't believe that he can cast a spell," Steve said.

"Just don't look into his eyes," Joey repeated.

The boats sailed into the open sea. Steve, wearing one of Joey's ceremonial shirts, his blond hair hidden under a conical Makah hat, felt himself part of the clan. His shoulders were covered with a Chilkat blanket that Grandma had lent him. It was woven of fibers of yellow cedar bark and it carried the images of the raven and the whale, symbols of the family heraldry. The animals were presented in a symmetrical split-in-the-middle form, as if they were spread out flat. The blanket had a faint scent, rather pleasant, reminding Steve of its origins within the core of the living tree.

"You be careful with my blanket," Cooing Dove admonished. "It was my dowry seven times ten years ago ..."

"I promise. I am honored that you allow me to wear it."

"She never lets *me* use it," Joey objected.

"You make fun of Grandma," the old woman said. "But the white boy respects your Grandma."

"You like him because he believes in Sasquatch," teased Joey.

"That too," she replied with dignity.

The colorful flotilla set its course along the coast of Washington, toward the tip of Cape Flattery. The travelers were to cross the Straight of Juan de Fuca to Vancouver Island and dock in one of its inlets. It was a trip of only two or three hours and Steve was glad that it wasn't longer. He already felt queasy; he was prone to seasickness.

To distract himself from the rising waves of nausea, Steve tried to concentrate on his surroundings. "I wish the Chief were wearing a war bonnet with hundreds of feathers cascading down his back, like in a Western movie," Steve thought. Instead, the Makah Chief's grass-woven headdress looked like a broad-brimmed Chinese hat with a narrow stovepipe of several levels attached to its top as if tin cans were stuck inside one another. Steve knew that the telescoping multilevels proclaimed the importance of the Chief. "Of course, different tribes have different types of dress, but just the same, I wish, Joey's dad had a war bonnet, like Sitting Bull," Steve thought. He turned his attention to the scenery.

He watched the rocky green coast of Washington slowly pass by on the starboard. A bluish, uneven mountainous contour of land began to loom ahead.

It was Vancouver Island, British Columbia, their destination.

V

Potlatch

They docked in a secluded inlet. The canoes were lifted off the decks of the fishing boats and lowered into the water. The people disembarked. At once they proceeded to paint one another's faces using theatrical makeup.

"Let me paint your face in our colors," Joey said. He made several vertical stripes of black and red, adding a few blue dots across Steve's cheeks.

"Hey, you look like a real Indian," Jim said as he decorated Joey's face.

Steve smiled. "Yeah, especially with my yellow hair!"

"Hide it under your hat," Joey advised.

Steve shook his head. "Nah. I'll be an "albino" Indian!" He watched the old people place rings in their noses, but Joey's grandmother was the only one to wear a *labret*.

The Chief waited for his people to finish their preparations.

He stood in the stern of the large leading canoe, his arms crossed over his chest. Suddenly, an imposing new figure wearing a Raven's mask rose on the bow of the Chief's canoe.

Steve looked closely, knowing that it was Jim, yet failing to recognize the familiar friendly features under the carved, ferocious-looking mask. Jim flapped his arms, and the feather contraption over his shoulders made him look as if he were a giant raven about to take flight.

The canoes took off, shooting out of the inlet like arrows, barely skimming the water. Paddling along with the rest of the people, Steve was free of nausea. The canoe rode over the water smoothly, without the pitch and the roll of larger vessels, leaving almost no wake.

The shaman resumed his song and the people picked up the chant. Steve, caught by the spirit of the moment tried to sing along. He was bursting with excitement.

Crowds of Nootkas met them at the water's edge. The two tribes greeted one another according to their rank: The Chief of the Nootkas greeted Joey's father; the future Chief embraced young Jim. Steve, as a special guest, was introduced to the head of the Nootkas, Howling Wolf, otherwise known as Mr. Tom Douglas.

"Howdy, boy, are you enjoying your vacation?" the Nootka Chief slapped Steve on his back.

"Yes, sir. I'm having a great time!"

"Good. Let me know if you want anything." Howling Wolf's attention switched to another group of canoes arriving at the pier.

"It's gonna be a great potlatch!" Joey beamed. "Here come the Haidas. All the way down from Alaska!"

Steve watched a dozen brightly decorated canoes slowly glide up to the pier. Each vessel had a different carved bow resembling the vicious profiles of predatory birds or beasts painted in strong colors.

"The Haidas have the most beautiful canoes in the world," Joey said. "We and the Nootkas are fishermen. But the Haidas are artists. Lots of their work are now in museums. They are famous for their totem poles. Anyway," Joey continued, his attention turned back to more practical matters, "my Dad and the Haida Chief will stay with Mr. Douglas. But we will camp in the meadow with the rest of the guests."

"Great! I prefer to camp."

Joey pulled his blanket closer to his body to exhibit its design for all to see. He led Steve toward the meadow where many campfires were already lighted.

The young people milling around the meadow all seemed to know one another. Joey swaggered among them introducing Steve as his best friend. "He knows movie stars!" Joey bragged. "His dad writes for TV!"

Steve grinned in embarrassment, thinking, "Big deal! I wish there were something *real* to brag about!"

"Let's go to the Chiefs' fire," Joey finally said. "All the action will be there." They trotted toward the central bonfire.

The fog came rolling in from the ocean clinging to the shore like sticky cotton candy. It tasted of kelp and smelled of sea creatures. Around the campfires the heat dissipated the fog, but beyond the perimeter of their light, clouds of gray mist closed in again. People emerged from it as nebulous forms, the mystique intensified by their masks which looked grotesque in the approaching twilight. Their voices sounded muted then suddenly exceedingly sharp, the fog muffling the sound.

Steve was seated with Joey's family. The other two Chiefs and their male relatives took their places around the fire as well, the woman sitting on the opposite side, according to the custom. Joey's grandmother, as the most senior person, was given a special seat of honor next to the wife of the host.

The Nootka Chief rose to his feet. He looked regal in his tribal costume with the carved mask of a wolf pushed back over his head. It created the impression that his head was clasped in the wolf's jaws.

"He's the boss of the *whole* Wolf clan," Joey whispered. "He's very important. See those ermine skins hanging from his headdress? They are worth hundreds of bucks."

Steve looked closer. He saw dozens of glossy white ermine skins cascading down the Chief's back and over his chest. The skins were whole, the tiny legs and even the claws of the ermine plainly visible. The skins rippled with every turn of the

Chief's body.

"Dear friends," Howling Wolf began, speaking through a microphone. "I welcome you to our shores. I'll speak in English because we have with us several friends who know no other tongue but English." The guests laughed.

"He means you." Joey poked Steve in the ribs.

"Nope. I know some French also," Steve said.

"Oh, *pardonez-moi*," Joey whispered making a prissy face, both boys guffawing.

Joey's father shot a warning glance toward them. They were being impolite.

"It's a great day for our tribe and for my family. We are celebrating my son's graduation from college," the Chief continued. "To commemorate the occasion, we'll have a totem pole dedication. Our good neighbors, the Haida carvers, made one for us. Then we'll have a potlatch. So, my friends, enjoy yourselves." Howling Wolf passed the microphone to Joey's father who thanked him on behalf of his clan. Then the microphone went on to the Haida Chief, Eagle of the Tall Trees. The short ceremony of welcome was over. The crowds broke away from the fire. They surrounded the long tables laden with cases of beer, soft drinks and urns of hot coffee as a delicious aroma of barbecued beef wafted through the fog still rolling thickly from the ocean.

Like two hungry puppies, the boys followed the aroma to the end of the meadow. Several cooking pits were set up on its edge and a crowd of spectators was already waiting for the first morsels of food.

"Here is where we stay," Joey said dropping down on the damp grass near the pit. The boys watched the Nootka women barbecue whole sides of beef and pork smothering the meat with thick, tangy sauce. Then another smell drifted toward them from yet another pit where the Nootka youths baked potatoes and corn raking them among the ashes with heavy garden rakes.

"All these smells drive me nuts!" Joey declared. "I must eat at once or I'll die!"

"This looks like one giant picnic," Steve said.

They ate grilled salmon and barbecued meats, blowing on their fingers burned on charred potatoes and corn pulled out of the ashes of the cooking pits. Steve had never tasted anything so good.

An urgent sound of drums summoned the boys back to the Chief's fire.

The clearing around the bonfire was enlarged by moving people further away. The three Chiefs, gathered together, waited to begin the ceremonial dance. Their relatives crowded behind them.

"What about you, Joey?" Steve said seeing that Jim and even Chickie were standing behind the Chief.

"Dad wanted me to stay with you."

"I'll be okay. Go ahead."

"Nope, Dad wants me to explain the ceremony to you."

Steve felt excitement mounting. The frenzied pounding of the drums reached its peak and then suddenly stopped. A hush fell over the crowd.

The three Chiefs entered the clearing. Slowly turning in one direction, then in another, they stood with their backs to the audience and their arms outstretched.

"They are showing the designs on their blankets," Joey whispered. "It's very important. The design on every blanket is different. It's like a business card, Dad says. It lets the others know who the person is."

The drums resumed their beat. The Chiefs began to move. They paused on every third step and the shamans, of all three tribes, seated together on the side, shook their rattles vigorously at every pause.

Having exhibited the splendor of their blankets, the Chiefs took their seats. Their wives and children entered now. They moved in a slow procession, repeating the same steps like in a dreamlike dance. It went on for a long time and Steve became restless.

"What's next?" he turned to Joey.

"Story dances. I'll explain the stories as they unfold. Our

tribe will be the first. We have one of the best dancers on the coast," he bragged.

"Who?"

"My cousin Jim!"

The long procession of the wives and children had finally ended and the people settled around the fire once more. The drums changed their beat to a new frenzied tempo. A dancer dressed as a Chief strutted to the center stage. He wore a carved mask with a human face surrounded with tufts of coarse hair.

"Real human hair. *Very* important for the story," Joey whispered.

Suddenly another dancer appeared. He wore the hideous mask of a snarling bear and his back was covered with a huge bearskin. He tottered around the fire, imitating a bear on its hind legs. Coming upon the Chief, he roared and struck the Chief down. Then, stealthily, he escaped into the darkness. The audience, mesmerized, did not clap, watching the story with total attention.

"The warriors will now discover the body," Joey commented. Sure enough, several dancers, Chickie among them, entered the space before the fire. They shook their spears decorated with feathers as they pranced around the fallen Chief.

A group of women entered, led by Sandy. They knelt at the prostrate body of the Chief, grieving. Then, they covered the body with a blanket and dragged it away by the legs.

"Now watch for Jim. He's terrific!"

The rhythmic beat of the drum abruptly stopped. In the sudden dramatic silence, Jim attired in his Raven's costume, leaped into the center before the great bonfire. His mask, with its huge beak, looked frightening in the changing light of the fire. Its eyes, made of iridescent abalone shells, reflected the flames, sparkling wildly, as if alive. Jim moved in nervous jerks, leaping and flapping his winged arms, strutting like a bird, twisting his head suddenly, then freezing for a moment as if listening. He was a superb dancer with a real sense of showmanship.

For the first time the audience burst into applause.

"The Raven orders people to punish the guilty bear. He says that the killer still has his victim's hair in his mouth," Joey commented in a whisper.

Jim, the Raven, glared malevolently at the crowd. Suddenly, the dancer impersonating the bear appeared from the darkness. The hideous beast held a bunch of dark human hair in its gaping jaws. The warriors pounced upon him, striking him down with their spears. The dancer made a great show of dying by writhing, leaping into the air and plunging headlong to the ground. He was rewarded with enthusiastic applause. Finally, the women swirled around him. They lifted the enormous bearskin showing the audience that the murderous beast was dead.

"Our people say that this is a true story," Joey said. " And Grandma swears that this is the original bearskin. Hundreds of years old! Can you believe it?"

"Sure. Anyway, the story and your commentary are fantastic!"

"I'm glad you liked it," Joey grinned, pleased. "That's nothing! Wait till the potlatch on the last day! Boy, you'll love it!"

VI

Joey's in Trouble

They slept in their sleeping bags around the fire which was tended throughout the night by the youths of the host tribe. In the morning, the sun shone brightly again although the lower meadows were still hazy, unable to liberate themselves from the grip of the dense fog of the previous day.

"The weather looks great for the competitions!" Joey said folding his sleeping bag.

"What kind of competitions?"

"The usual. Baseball, track, spear throwing and archery."

"Archery?" Steve's interest perked up.

"Yep. Unfortunately, last year we were beaten by the Haidas. They trained the best archers you've ever seen. They were far superior to our team. I am afraid we might lose again. I see the same kids are back this year."

"I am an archer myself," Steve said. "I'm on our county team."

"No kidding? Why didn't you tell me before?"

"It never occurred to me."

"After we get home, let's go and shoot a few arrows... Hey, how would you like to see the totem pole before it's put up? I know where it is," Joey said changing the subject.

"Sure."

They ran to the road at the end of the village. A huge truck was parked at the shoulder. Chained to its long open platform trailer was an enormous carved log.

"A sleeping giant," Steve thought, touching the chains which crossed the totem's broad chest like the bonds of Prometheus.

The totem was carved out of a single tree by a Haida master carver and two apprentices. It was more than sixty feet long and at least eight feet in diameter. At the top of the totem was a figure wearing a wolf's mask, but dressed in human attire.

"This is the subject of the pole: the Chief's son, the graduate," Joey said. The figure was painted in strong primary colors, its feet firmly planted over the shoulders of a wolf.

"His ancestor," Joey pointed to the wolf. "Do you think you can read the rest of the totem?"

"I'll try." It was difficult to follow the design. One symbol melted into another, the leg of one animal becoming the body of another. Out of each carved eye and even out of every claw, there peeked faces of frogs or minks or foxes.

"I give up!" Steve laughed. "I can recognize only one other image—a Raven. It has a straight beak in contrast to an Eagle which would have a curved one. Am I right?"

"You sure are. But, don't feel bad. Even I can hardly read the totem. I think only my grandma would understand every detail of it," Joey said.

Several men in hard hats approached the truck. It was time to move the totem to its place in front of the Chief's house. The boys climbed down from the platform.

The truck moved slowly toward the center of the village jammed with spectators where a tall crane waited to lift the sculpture off the trailer and place it in a specially prepared deep hole. Amidst much shouting and the noise of the grinding gears, the totem was lifted. The men in hard hats guided the unloading as if it were a laminated beam in a tract house subdivision.

Steve felt deflated. His romantic imagination demanded

hundreds of bronzed, seminaked natives pulling and pushing, placing the totem in its place amidst chanting and the beating of a drum. The grinding of a crane, the workers in their hard hats, somehow looked mundane and out of place, Steve thought.

Suddenly he had an impulse to write to his parents. He missed them. Not a lot, but he missed them. He was bursting with new impressions, and he wished his father were there to share them. He made his way to the community center and, finding a quiet corner, sat down to write them a letter.

On the day of the archery competition, Joey awakened with a severe abdominal pain. He vomited and later came down with diarrhea. He felt weak, his face covered with cold perspiration. "How can I compete when I must run to the toilet every five minutes," he lamented.

His teammates were unsympathetic.

"You should've thought of that before you pigged out yesterday, gobbling everything in sight," the team captain Dan Beaver Tooth reproached him. "Have you seen the Haida team? Those guys are tough!"

"Big deal!" Chickie snorted.

"You won't say that if we lose," Dan snapped. "We are already one man short but with Joey out we'll be two men short."

"Who cares," Chickie swaggered, spitting at the sand. "I can beat them blindfolded."

"You could, at one time, but no more. You have been skipping practice. The Haida guys worked as a team all year long."

Chickie glowered, but said nothing more.

"The Haida team brought along a new kid. He's an Alaskan junior champion," Dan turned to his teammates.

"It ain't fair. He ain't even of the Eagle clan," Chickie objected.

"He's the guest of the clan, and he becomes one of them for the period of his visit. It's the tribal law and you know it," Dan said.

The team contemplated gloomily the news about the new

archer. Steve could see that the boys were preparing themselves for the inevitable defeat.

"Does this law apply to all the guests of the tribe? Even to the non-Native guests?" he asked Joey.

"What are you driving at?"

"Well..." Steve hesitated, and then plunged in. "As I've told you, I am an archer myself. I am on the Los Angeles County Archery Team. I placed second in the California Junior Spring Meet this year. Would you let me shoot on your team?"

"Right on!" Joey shouted. "Now we'll have a full team!"

"No!" Chickie growled. "We don't need no White Bellies on our team. He's lying that he can shoot! No White Belly can be an archer!"

Joey stared at him with fury, but controlled himself and said, trying to sound calm, "We still have three hours, why don't we test him?"

"I'm willing. Give me a bow...and a couple of arrows," Steve said.

Dan Beaver Tooth handed him his bow. It was made of smooth yew wood with its ends carved upwards. Steve tried the bowstring made of seal gut sinew and it reverberated at the touch of his fingers.

"Cool!" Steve could hear the bowstring sing. "I've never seen such a terrific recurve bow! Let me see the arrows."

Dan pulled an arrow out of the quiver behind his shoulder. Steve balanced it on his finger. The slender arrow was beautifully rubbed, its shaft satiny to his touch. The sharp tip was crowned with a splintered bone, while the blunt end was fletched with double sections of owl feathers.

"Terrific!" Steve exclaimed. "Our arrows usually have plastic vanes instead of feathers. You guys really know how to make fantastic arrows!"

"Cut out the crap and shoot," Chickie yelled.

"Pay no attention to him, take your time," Joey said, cramps seizing his abdomen again, doubling him in pain.

Steve glanced at him uneasily, but Joey smiled weakly, murmuring, "I'll be okay, don't worry."

Steve set the arrow against the nocking point on the bow. He spaced his feet at shoulder width, stretched his left arm and slowly pulled back his right one.

"I'll shoot at that stump," he said. A large chunk of rotted driftwood was fifty or sixty feet away, partly obscured by sand drifts brought in by the tides.

"Bid deal!" Chickie snorted. "It's as big as a house!"

"Let him alone!" Beaver Tooth snapped.

"Take it easy," Steve said to Chickie. "You wanted to see if I can shoot. Okay, I'm demonstrating." He drew the bowstring with his three fingers, and gently released it. The arrow flew through the air hitting the stump almost instantly. "Now give me another arrow and I'll place it next to the first one," he said.

Dan handed him another arrow. They all knew that it required a well-trained archer to "group" two arrows in a row, especially when there were no guiding lines and colors of the concentric circles of the target.

Steve sensed the boys' apprehension. He knew that the real test of his ability rested with the next shot. He nocked the arrow, feeling the tension of the bowstring. He had a premonition that he would place the arrow exactly next to the one still quivering in the stump.

He released the arrow and it landed less than an inch from the first one.

"Terrific!" Dan slapped Steve on his back. "You're in!"

"Thanks. But I would like Billy to agree. I would like it to be unanimous." Steve had deliberately called Chickie by his given name. "Well?" he pressed.

"Oh, what do I care, okay with me," Chickie finally grumbled, spitting at his feet, as was his habit.

"We'd better coach you about our rules," Dan said. "It's really very simple," he continued removing the arrows from the stump and checking their tips for damage. "We move along a beach or the woods and shoot at targets from different distances. Our targets are usually stationary, but there are always some surprises, half hidden. It's all quite simple once you get the knack."

"Ain't you scared, White Belly, to shoot against the best archers in the Northwest? Just don't wet your pants when you come up against the Haidas," Chickie taunted him, stubbornly refusing to give up.

"If you don't shut your trap I'll bring you before the Tribal Council," Dan threatened.

"We don't have much time, guys," Joey said. "We must paint our bodies."

"Sure, paint a pretty flower on his white belly," Chickie sneered.

Steve felt anger rising in him. "Lay off, Billy," he said.

"Yeah? You wanna fight?"

"No, I see no good reason to fight."

"Coward!"

Steve swallowed hard, his fists closing tight until his nails dug deep into his flesh. He wished he could charge at Chickie, pounding at his loathsome face. But he controlled himself. Trying to keep his voice calm, he said, "I'll fight you by proving that I am a better archer."

"Okay, guys, let's get going" Dan called. He opened a small tin of theatrical makeup.

The boys stripped to the waist. They painted one another's faces and torsos with stripes and circles, Joey decorating Steve's face and body.

Meanwhile, two other groups of archers appeared on the beach. They were also stripped to the waist, their faces and bodies painted in bright colors.

The competing captains huddled together to discuss the rules of the competition, including Steve's participation on the team.

The broad beach quickly filled up with spectators. A small pickup truck appeared. Riding near the water's edge, its tire marks

erased by the gently lapping waves as fast as they were imprinted on the wet sand. It carried the targets and the bales of hay to be placed behind the targets for safety.

Steve felt his mouth turn dry. He swallowed hard, fully aware of the importance of his performance for Joey's team. But the premonition that he was destined to do well was still with him, surrounding him like an invisible cloak.

Dan Beaver Tooth returned to his teammates. "The captains have agreed to Steve's presence on our team. They have also suggested that we tie whalers' knots."

"It's a special hairdo the whalers used to wear in the old days," Joey explained to Steve. "It used to be part of the ceremony of the hunt, so the guys want to keep that tradition going. Besides, it's practical. It keeps hair out of your eyes."

Dan passed several seal gut strips to his teammates.

"Let me do it for you," Joey said. "Sit down, you're too tall for me, I can't reach," Steve sat on the sand. Joey gathered Steve's long hair in one hand, and pulled it up to the top of his head. Then he wrapped the leather string around it, tying one knot. Folding the resulting ponytail in the middle, he tied it once again, creating a rounded topknot at the peak of Steve's head. "Now you look like a real Native American!"

A Nootka tribal official, wearing an armband for identification, approached the boys.

"I'm your field captain," he introduced himself. "Let's draw." He held three sticks of different lengths in his closed fist and the captains each drew one. The Haidas pulled the longest stick. They were to start the competition. The Makahs were next and the Nootkas last.

The teams faced one another in a semicircle, waiting for the signal to begin. They saluted each other by bending their right arm at the elbow and raising their hand to shoulder level, with an open palm facing the opponent, Steve copying every move of his teammates.

"Here we go..." Dan murmured. "The first distance is thirty

yards. Just keep your cool."

From the corner of his eye, Steve saw Joey grab at his abdomen. His face was ashen even under the gaudy makeup.

Steve touched Joey's hand. It was cold and clammy. "Shouldn't you go to the first aid station?" he said quietly.

"Nope. I'll be all right."

Steve felt waves of anxiety flowing from one teammate to another. He looked closely at the competitors, trying to guess which one would be the most difficult to beat. They were all of the same age—thirteen and fourteen. He and Dan were the tallest, with the longest arms, a possible advantage for their team, but otherwise they all seemed to be well-matched.

The field captain climbed up on the platform of the truck.

"Friends, may I have your attention, please," he announced through a bullhorn. "There will be *individual appraisals* of the contestants during this meet. At the end of the competition, we'll select the best archers among all three teams to train for the international competition, for a chance to be on the All-Native American Team at the next Olympics."

"Oh boy, I'll never make it," Joey whispered, his face distorted by pain.

"Don't say that! You've got to keep trying!" Dan hissed fiercely.

The Field Captain signaled and six targets were placed in front of bales of hay. He blew his whistle to clear the field.

"Get ready!" he shouted. "Attach your identifications! Good luck!"

Steve pinned a piece of canvas with his identification number, Makah-6, to the right hip pocket of his jeans and helped Joey with his.

Joey's face was covered with sweat. He was obviously in pain, but was determined to go on with the competition.

"A fine time to have a bellyache," Chickie grumbled.

"Shut up, Chickie." Steve was through with being polite. "Joey is sick. He ought to go to the hospital. It could be something serious...maybe appendicitis...."

"What are you—a doctor?" Chickie sneered.

"I had appendicitis myself and Joey's symptoms are very much like what I had."

"Yeah, I think I had better go to the first aid station...I must lie down." Joey's lips were drained of color.

"I'll take you there. Hold on to me," Steve said.

"No. You stay here. The meet is about to start. Here, take my equipment." Joey passed his bow and arrows to Steve. He tried to smile but the pain was too severe. He reeled off his feet, and doubled up on the ground.

Even Chickie became alarmed. "Hey, we'd better get him out of here," he said nervously.

"Call Chief Flying Raven!" Dan ordered one of the children from the crowd of surrounding spectators. The boy took off like a hound in pursuit of a rabbit.

The field captain jumped down from the truck. "What happened?"

"The Little Raven is sick," Dan said.

Steve knelt on the ground next to Joey. He wiped perspiration from his brow, smudging the bright paints on his face. Joey's eyes kept rolling up, exposing the whites, his body burning hot now. He lay on the sand, his hands pressed to his rigid abdomen, his lips stubbornly shut.

The spectators surrounded the team.

"He is a real brave!" Steve heard someone say with respect. "He suffers pain nobly!"

The crowd parted and Joey's father stepped in.

"What happened to Little Raven?" he said in alarm, kneeling on the sand next to his son.

"He's sick," Dan said quietly.

The Chief picked up Joey in his arms. "What's wrong, son?" he asked tenderly. "Someone, get the truck! I must take him to the hospital! Quick!" the Chief commanded. Several boys dashed toward the pickup truck.

"May I go with you, sir?" Steve said as the Chief climbed into the cabin of the truck cradling Joey in his arms.

"Tell him to stay with the team," Joey said weakly. "He must take my place..."

"You heard him," the Chief said.

"We count on you." Joey closed his eyes as another cramp seized his body in its grip.

The truck spun its wheels, tossing sheets of tiny grains of sand into the air. As the wheels found purchase, it lurched ahead at full speed.

The Chief and Joey were gone. The archers of the three teams mingled together, Joey's illness removing the competitive feeling among them.

Chickie remained aloof. He hated Steve more than ever now. He was unable to accept Steve as an equal member of the team. He wished that Steve would lose his nerve and fail at the competition. Even if it meant defeat for the team.

"Okay, people, let's get back to our business," the Field Captain shouted. "Address the targets!"

Six Haida boys stepped forward. They lined up facing the targets thirty yards away and raised their bows.

"Shoot!"

The arrows soared into the air and Steve heard them vibrate as they hit the targets almost simultaneously.

"Reload!" The Haida boys nocked new arrows.

"Shoot!" Again, the arrows rose and hit the targets. All of them were within the gold confines of the targets.

"They are terrific, these Haida kids," Steve thought. "It will be tough to battle them for points. *I must hit only the gold!*" he thought with emphasis, psyching himself up by repeating it several times.

The officials and the captains of the teams approached the targets. They marked the scores for each contestant on score cards and withdrew the arrows from the targets.

The Makahs were next.

VII

Archery Competition

The boys lined up before the targets. There were only five of them now, a disadvantage for their overall score.

Steve felt a stir of anxiety at the pit of his stomach, like before an exam, but despite it, he felt that nothing would go wrong for him that day. "I am going to win!" he kept telling himself.

"Address the targets!" commanded the field captain. "Shoot!"

The arrows winged toward the targets. Steve, Dan and Chickie all hit the center of the targets in the gold. Nine for each of them. However, the arrows of Fred and George, the twins, were stuck in the hay. Two zeros. It was not a good showing against the perfect scores of the Haidas.

Steve felt the eyes of hundreds of people on him, but it did not bother him. He felt that he was invincible today.

"Address the targets!" Gently, Steve pulled on the bowstring, aiming just below the center of the target, adjusting for the short distance.

"Shoot!" The arrow rose swiftly, hitting the gold. There was a scattering of applause from the spectators.

"Another nine!" Dan yelled, slapping Steve on his back, forgetting the decorum of being the team captain and his own disappointment at hitting only a seven.

"Just lucky," scowled Chickie, not hiding his envy. He was able to hit only a five.

"No, just superior!" Dan corrected him.

The Nootkas were next. Steve watched them closely, but only two boys managed to score nines. However, it was still too early in the meet to write anyone off.

The next target was set up in the brush, the targets partially obscured by small trees.

The Haida archers again did very well. They were well-trained, very much like Steve's own California team.

"The Haidas are the ones to beat," he thought. "I must hit only the nines!" he told himself. "God, dear God, *I'll never ask for anything again in my whole life, let me hit only the nines!* Please, dear God!" he prayed.

Dan and Steve both finished the set gaining nine points for each of their arrows. Chickie was erratic, making a nine and then sending an arrow into the bale of hay. The twins managed two nines and two sevens.

"It's much more difficult than a classic meet," Steve thought, but his nervous energy acted like a life jacket, buoying him up. It was *his* day. He knew it.

The Haida boy, Tommy Sharp Claw, and Steve were the only archers who were constantly credited with perfect scores.

"It stinks! I feel the evil spirits around me!" Chickie muttered, spitting in Steve's direction.

The teams, succeeding one another, moved steadily across the wooded area facing the targets at different ranges. Some of the targets were at the ground level while others were hoisted among the trees, moving with the wind, requiring constant adjustments in aim.

Steve's perfect scores drew admiration from the spectators who cheered his every hit.

"This white kid is something else," he heard someone say.

Finally the meet was over. The teams saluted each other by clasping their hands against one another's shoulders in Roman style. The archers were immediately surrounded by crowds of enthusiastic admirers while they waited the official announcement of the scores.

As everyone had expected, the Haida team had won. The Makahs took second place; the Nootkas last.

"We would have been last had it not been for Steve," Dan said to his teammates. The boys nodded in agreement.

"Speak for yourself," Chickie snarled.

"Shut up! You're a disgrace to our team. I propose that we expel you!" Dan turned to him angrily.

Chickie spat at the sand, and turned his back on his teammates.

In the evening the archers were presented to the tribes. Wearing their face paints, with their hair still gathered in the whalers' knots, the boys danced before the Chiefs, displaying their bows and arrows. Steve danced also, copying the movements of the other boys.

Chief Flying Raven was back at his place of honor. Steve was anxious to inquire about Joey but did not dare to approach him during the ceremony. He knew that Joey had been taken to the hospital by a helicopter. He had heard the machine hover over the campgrounds. Seeing Joey's father back among the guests was reassuring. "He wouldn't be here if Joey were in danger," Steve thought, relaxing.

"...We have several outstanding archers," the field captain was saying over the microphone. "One of them, Steven Bradley, is not eligible to train as one of us. It is unfortunate. But we honor him for his superb marksmanship. He couldn't have done better if he were a Makah, a Haida or a Nootka! Steven Bradley, step forward, please!"

Steve felt blood rush to his face as he stepped out of the ranks.

"Here is your trophy. As a guest of the Makah clan, you deserve to be the first to be congratulated!" He handed Steve a gilded statuette of an archer attached to a pedestal made of imitation marble.

"Thank you, I am honored." Steve shook the Captain's hand.

Next to be acknowledged was the Haida boy, Tommy Sharp Claw. He was handed an identical trophy as were Dan and three other boys from the Nootka and the Haida clans. Chickie won nothing.

"We hope that you boys will be among the competitors at the next Olympics," the Captain continued. "Good luck!" Everyone whistled and stomped their feet as the winners stepped back.

Steve felt someone tugging at his arm. He turned around. Joey's grandmother was crouching behind him.

"You good boy," she mumbled, the *labret* making her speech almost unintelligible. "You shoot good like real brave. I watch you," she smiled. "Little Raven in big city. Victoria."

"Is he all right?"

"Sure. He is full of courage. White doctors cut out bad part from his belly. Bad spirits gone. I talk to him on the telephone."

Steve grinned. "Good." He knew that Grandma never talked on the telephone. She believed that the invisible wicked spirits were waiting at the other end of the line to snatch her voice away.

"I talk to Joey. I tell him you won at archery. I tell him you good friend," she repeated.

"Thank you, Grandma. Can I talk to him?"

"No. Only his father and his grandma."

"Tell him to get well fast."

"I know what to tell," she said withdrawing back into the crowd.

"Who's with him?" Steve managed to ask.

"Jim and Sandy." She was gone.

"Good," Steve thought. "At least he's not alone. It's important not to be alone." He recalled how he himself had been taken to the hospital some years ago. He and his mother had been at the Lake Arrowhead resort, his father staying at home working on a TV script, as usual. The attack of appendicitis had hit Steve with sudden savagery, but his mother recognized the symptoms and had taken him to the hospital in Los Angeles without waiting for an ambulance. He remembered how she had driven along the twisting mountain roads, holding his head on her lap with her right hand while steering the car with her left. She kept stroking his forehead, wiping the sweat off his face with her soft fingertips. She guided her car expertly, manipulating the hairpin turns with the ease of a racetrack driver. She was stopped on the freeway for speeding, but once the patrolman saw the emergency, he escorted the Bradleys to the hospital, his siren wailing and the lights flashing. All through the ordeal of the mad ride, Steve's mother pressed him to her side tightly, and he remembered what a great feeling of comfort it gave him, despite his excruciating pain. After the surgery, his mother's face was the first to greet him as he regained his consciousness. He was surprised that he still remembered this episode so vividly.

"Too bad that Joey will miss the potlatch," he heard Dan Beaver Tooth say. "He was so looking forward to it."

The host, Howling Wolf, rose from his seat, the gleaming white ermine skins of his headdress reflecting the lights of the great bonfire.

"Friends and guests," he began, his voice amplified by a loudspeaker. "We are coming to the end of our celebration. Tonight we'll have our potlatch. I want to thank the members of my clan for the goods contributed so generously for distribution. Long ago potlatches were held to brag about one's wealth. Gifts were expected to be returned with interest. We don't do that anymore. Times have changed. Keep your presents. The Wolf clan expects no interest on its gifts. However," he paused, "if you want to reciprocate with a potlatch of your own at some future date, rest assured we'll all be there!" Everyone laughed. The Chief Howling Wolf continued, "Will my sons and my nephews step forward to distribute the gifts, please." Several young men gathered around the Chief.

John Timber Wolf, the future Chief of the tribe, picked up two large identical cartons and passed them on to his uncle, the Chief of the Nootkas.

"For my esteemed brothers, Flying Raven and Eagle of the Tall Trees," Howling Wolf handed the gifts to the Chiefs. They lifted the heavy cartons above their heads to show the crowd that they contained portable color TV sets.

The nephews and the sons of Howling Wolf dashed about like a platoon of shipping clerks, their wolf masks pushed off their faces, their Chilkat blankets trailing behind them.

Soon all the packages were parceled out, and the guests felt free to open them. The wrapping paper and packing materials were thrown into the fire as the gifts began to emerge.

Instamatic cameras, transistor radios, pocket calculators and watches, portable sewing machines and electric coffee makers — all were among the gifts of the generous Nootkas. The children were given toys and clothing.

"And not a single Chilkat blanket," Steve thought, remembering reading about Indian potlatches of bygone years. The most valuable items in those days were the Chilkat blankets. The tribe's wealth was judged by the number of blankets changing hands. "Aren't you sorry that they don't distribute blankets anymore?" he turned to Dan.

"Nope. Who needs them? My mom has a trunkful of blankets. I'd rather have a camera, or a good transistor radio."

The speeches began. The heads of the families rose up one after another, thanking the Chief Howling Wolf and his tribe for the generous gifts.

Steve became bored. He thought of slipping away unobtrusively when he heard his name spoken. John Timber Wolf stood behind him.

"We are honored that you took part in our games. My uncle

wants you to have this," he said. He handed Steve a large recurve bow and a quiver made of animal hide and brisling with arrows fletched with colorful feathers.

"Gee...awesome..." Steve stammered, suddenly unable to find the right words. "...Thanks..."

The young man grinned. "I knew you'd like it! The best bow maker on the Vancouver Island made it. An identical set is in the B.C. Provincial Museum in Victoria."

"...Oh, wow! Gee thanks...cool."

"Enjoy it! You're a champ. I wish I could shoot as well as you do!" John was gone.

Steve ran his hand over the smooth curve of the bow. "A real Indian bow! Something I've always wanted!" he exclaimed turning to his teammates.

The boys crowded around him admiring the bow. None of them had ever possessed such a fine set. They generously acknowledged that Steve deserved his magnificent present.

Only Chickie refused to congratulate Steve. He pretended to be disinterested.

VIII

Journey Back

It was getting dark. The night stalked like a black cougar of native legend, right outside the rings of brightness caused by the campfires. The Chiefs wearing their full regalia and carved masks, and the guests draped in their colorful blankets, created a dreamlike tableau.

"I wish Dad were here!" Steve thought. "What a scene for a screenplay! Better than anything that I've ever read, or seen in the movies! And it's for real!"

The persistent beat of a drum silenced the din of people's voices. "What now?" Steve turned to Dan who in Joey's absence became his guide.

"The Nootka shaman is going to dance."

"Great! I've always wanted to see a shaman perform. Will you explain what he'll be doing?"

"Sure. But I am not too good at this old stuff. I'm a modern man." Dan looked sheepish.

The Nootka shaman bounced before the fire. He was naked except for a narrow loincloth and a wolf's skin over his shoulders held together by the animal's paws. His entire body was tattooed in intricate swirls descending to his ankles. Like the Makah shaman's, his hair was long and matted as it stuck out from under his wolf's mask.

The shaman grabbed a burning branch from the fire. Waving it back and forth, he made patterns of smoke that remained suspended in the air creating the illusion that the shaman was encircled by the coils of a vaporous boa constrictor. The spirals looped around him, the shaman disappearing among them as if he had been swallowed by the monster, only to reappear again.

"Boy, this is weird!" Steve whispered.

"He's trying to scare people. Are you scared?"

"Well...not really. But Joey said that a shaman has an evil eye? Do you believe it?"

"I sure do. Never look directly into a shaman's eyes!"

Steve felt a shiver run down his spine. He thought of the hypnotizing stare of the Makah shaman. "It's just a bunch of superstitions," he told himself, but the feeling of apprehension remained.

The Nootka shaman's ferocious mask of a snarling wolf, which hid his face, suddenly sprung open in the center, like saloon doors in a Western movie, revealing another mask, a horribly distorted human face.

The people shrieked even though they had seen such stage effects before. But for Steve, the sudden appearance of the mask was shocking. "Boy, he startled me!" He laughed nervously.

"Me, too," Dan said. "It doesn't matter how many times I have seen it—it always scares me! But the shamans have even more cool things. They make special puppets that fly on hidden wires. They look like shrunken people, just horrible! The shamans rig a system of pulleys so that the puppets move their arms and legs and even fly across the room. Some people believe that they are the spirits of the dead," Dan said in a low voice.

"Do you believe it?"

"Of course not. But I used to be scared to death. When I was a kid, our shaman had another trick. He had a system of kelp stems fixed under the mats in the tribal house. He would chant and his voice would come out somewhere else, like from another world. It used to scare the pants off me! But then, some kids discovered the hidden kelp stems, which are like hollow tubes. They traced them to the corner from which we heard the voice. Then we all knew that the old coot was fooling us. But of course we told no one about it. The shaman has lots of *real* magic power, so we kept our secret."

"Quite a story."

Finally, the shaman slowed down his gyrations. The people joined him in a dance, all three tribes mixing together.

Steve looked for Joey's father, but the Chief was gone.

The potlatch was over.

It was foggy and cold next morning as Steve climbed out of his cozy sleeping bag. He was determined to seek out Joey's father.

"I must find out about Joey before we leave," he said to Dan. He headed toward Chief Howling Wolf's house.

He paused in front of the great new totem. It looked forlorn, its top figure lost in a shroud of fog as it curled in thick layers at its feet.

He found the Chiefs having breakfast around the large formica table in the kitchen. The Chief's wife was making pancakes, and the kitchen was full of the delicious aromas of freshly brewed coffee and fried bacon. Steve became acutely aware that he had not yet had his breakfast.

"Sorry to disturb you but I wanted to find out about my friend Little Raven," he said to the Chiefs.

"Ah, the hero of the archery competition! Come in, have some pancakes and bacon," Howling Wolf greeted him. "Margaret, set up another plate for our young friend here." The stout woman at the stove turned around, her face shiny from the heat of the griddle.

"I watched you yesterday. You were terrific! Super!" she smiled and two dimples appeared on her round face.

"Thank you." Steve realized that he was praised so exorbitantly because he was a guest. Were he one of their own, they would have accepted his prowess as hardly worthy of mentioning.

It would have been expected of him.

Chief Flying Raven smiled. "Joey's doing fine. I called the hospital just a few minutes ago. They performed an emergency appendectomy yesterday and this morning he's already sitting up in bed. Jim and Sandy are with him now. I am on my way there myself."

"Sit down, Steve, have some breakfast," the host insisted. "Bacon or sausage?"

"Both," Steve said boldly.

"Atta boy! Have some coffee, too!" The Chief poured steaming coffee into a mug and pushed it toward Steve. "Or would you rather have milk?"

"Milk, please."

"I hear you did very well yesterday," Joey's father looked at Steve with parental pride. "I am going to call your dad. Not only about your archery winning. I'm going to tell your dad how much we enjoy having you with us! You're a true friend!"

Steve blushed, the tips of his ears burning as if they were on fire.

"I don't know what to say..." he blurted out. Then, correcting himself, he said, "Yes, I do. I want to thank Chief Howling Wolf for the beautiful bow and arrows that he gave me at the potlatch. I've never seen such a terrific set! Also, I want to thank all of you for inviting me to spend some time among you. I'll never forget this summer. It's been the best summer in my life!"

"We're glad that you enjoyed yourself. As for the bow and arrows set—you deserved it. Use it in good health," Howling Wolf smiled. "So, when do you need the car to go to Victoria?" he turned to Joey's father resuming the conversation interrupted by Steve's arrival.

"Right away. Jim will bring it back. I'll stay with Joey until his release from the hospital."

"Okay. We'd better start cracking, but let's finish our breakfast. Sit down, Steve."

Jim and Sandy arrived from Victoria several hours later.

The clan at once began to pack its gear. Having made their symbolic journey into their past, the people were now ready to return to the reality of their present lives.

"Joey sends his greetings and congratulations," Jim said brusquely to Steve. Then he turned to the chores of launching the Chief's great canoe.

"Thanks." Steve became accustomed to the Native habit of wasting few words.

The canoes were loaded with the potlatch gifts and bundles of ceremonial clothing and masks. The people gathered on the landing dressed in jeans and windbreakers, the men wearing hip-high fisherman's boots and the women's hair tied in kerchiefs. They exchanged good-byes with the members of the Nootka clan, the women kissing one another on the cheeks. "Just like mom's friends in Beverly Hills," Steve thought.

Jim paced the pier waiting for the host, Chief Howling Wolf. He was impatient to leave. He hated to sail at night with so many children aboard the overloaded boats.

Finally, the Nootka Chief appeared. He had changed into a business suit, and, except for his hair plaited into two braids, no longer resembled the Indian Chief. "He looks like a candidate for political office," Steve thought.

"Good-bye, sir. On behalf of my uncle Flying Raven and the entire Makah clan, I thank you for your hospitality," Jim said shaking his hand. "We enjoyed the celebration and appreciated your potlatch gifts. Our boats may capsize, so heavy are they laden with your precious gifts!"

Howling Wolf looked pleased. "May your clan prosper," he replied. "Have a safe journey home."

Jim stepped into the leading canoe. As it glided away from the landing, the Makah shaman began to chant. Steve glanced at him over his shoulder. The shaman was staring directly at him. Their eyes met and locked for an intense second.

Steve shuddered. He had disobeyed Joey's warning. He had looked into the fiercely penetrating eyes of the witch doctor.

The palms of Steve's hands became moist. The paddle slipped out of his grip and drifted away.

"The evil eye.... It has already started," Steve thought.

The Makahs arrived at the inlet. Everyone pitched in reloading the goods from the canoes and hoisting them aboard the fishing vessels, hurrying to be on their way home.

The wind was increasing. Within moments dark clouds covered the sky as the wind whipped up the waves into roaring breakers.

"I lost a paddle," Steve confessed to Dan who rode in a canoe behind him.

"No you didn't. One of the men in my canoe fished it out. You can get it from him."

"Let's go, people," Jim shouted. "The storm's coming!"

"See you at home," Dan said getting aboard his father's boat.

Climbing aboard Jim's boat, Steve immediately became aware of the rising waves of nausea. In the rough waters of the Strait of Juan de Fuca, he knew that momentarily he would be seasick. He pushed his way through to the port side. And none too soon. Bending over the bulwarks, he retched, starting a cycle of misery. He thought of the comfortable bunk in Joey's room, where the floor was not constantly moving under his feet, where no cutting wind penetrated his clothing, where no pitching and falling horizon played tricks on his vision. He wished he were there.

Three hours later, pale and exhausted, Steve wobbled ashore at the Makah landing.

Someone touched his shoulder. He turned around.

"You forgot your archery set," the shaman said handing him his bow and arrows.

Steve lowered his eyes but it was too late. He was gazing straight into the black slit eyes of the smiling witch doctor.

"Thank you," Steve mumbled. The shaman stepped aside and disappeared among the disembarking villagers.

Steve was frightened. Twice, he unwittingly challenged

the native belief. A feeling of foreboding enveloped him, anticipation of something dreadful about to happen.

Steve staggered up the hill toward the Chief's house. The dogs greeted him, prancing around. He weakly protected his face from their exuberant wet tongues while trying to open the door.

Joey's grandmother was moving around the house already dressed in her usual clothes. Her *labret* was gone. "How did she get here so fast?" Steve thought.

"I come yesterday," she said, answering his mute question. "I know Chief in Victoria with Joey. I take care of you."

"But how did you get here?"

"I paddle my canoe."

"By yourself? Alone?"

"Sure. I know how to paddle canoe."

"She's terrific!" Steve thought. "She paddled across the choppy waters for several hours all by herself to come home in time to take care of me!"

He wanted to hug Cooing Dove, as he would have hugged his own grandmother, but he felt shy.

"It's kind of you," he said, taking her bony hand in his.

She waved him away like a pesky fly. "Go rest. You look green, I give you bitter root tea. You feel good soon."

Steve went to Joey's room and threw himself on his bunk. "A woman more than eighty years old, paddling a canoe for a hundred miles of stormy sea...incredible!"

He closed his eyes. He felt the room and the bed swing back and forth under him. He made an effort to concentrate on something else rather than his heaving stomach and burning esophagus.

"Sasquatch...I'll think of Sasquatch," he told himself. The image of Sasquatch, as he was generally portrayed by artists, filled Steve's mind. He saw a huge, hairy creature loping across a clearing as he had seen it on television in a documentary. The creature came toward him, and looking directly into his eyes, whispered, "You forgot your archery set." Steve was asleep.

IX

Dr. Stone's Lab

When Steve awoke three hours later, night had already fallen on the reservation. He heard the distant forlorn howl of a coyote to which the young dogs in the village answered with short bursts of hysterical barking. The older dogs did not bother to bark. They knew all the night sounds of the forest. They lifted their ears for a few moments, listening, then curled up into even tighter balls, sighing contentedly, secure in belonging with the people.

The bright, cold moonlight engulfed the small room. Steve shivered and reached for the covers. He heard footsteps and a slight creak of the door as it was opened by Joey's grandmother.

"I'm awake," he greeted her.

"You feel better?"

"Much better. All I needed was a little sleep."

"Good. Drink this. It taste bad but is good for you. No more bellyache." She handed him a coffee mug filled with warm liquid. "Drink!"

Steve tasted the brew with the tip of his tongue. "It tastes awful!"

"Drink!" she commanded and turned the light on to make sure that Steve would not try to fool her. She sat at his feet on the bunk and smiled at him, her flat face becoming a map crossed with hundreds of deep lines as if they were roads and canyons.

"But my stomach doesn't hurt anymore," he stalled.

"Drink!"

"Oh, okay," Steve sighed with resignation and swallowed the bitter liquid in several large gulps, without pausing for breath.

"Good. Now you feel fine! You good boy. Good archer. I watch you shoot," she said. "Sleep more now..." she shuffled to the door and closed it gently.

It was not quite six o'clock in the morning and the village was still wrapped in a shroud of fog when Steve heard Lucy's voice in the kitchen. The dogs barked and growled on the porch as they reluctantly admitted Lucy's dog, Tiny, onto their territory.

Steve quickly pulled on his pants and sweater. Without washing up, he rushed to the kitchen. Lucy, in the kitchen!

She was already seated at the table, her dog filling up the space under it. Joey's grandmother was stirring oatmeal in a large pot. The kitchen was full of warm smells of morning, of coffee perking on the counter and bacon sizzling on the stove. It made Steve's stomach contract, reminding him that it was empty.

"Hi!" he greeted Lucy. "Good morning, Grandma!"

"Grandma invited me for breakfast," Lucy said

"Great!" Steve couldn't think of anything to say. "What a jerk!" he thought, disgusted with himself.

"Sit down to eat!" Grandma ordered without turning from the stove.

"There was some news from Dr. Stone while you were gone. They found fresh footprints and they made plaster casts. They came out perfect!" Lucy said.

"Did they *find* Sasquatch?" Steve's shyness was gone.

"Dave Rosen photographed some creature with his telescopic lens. But he said it was hard to say what it was. It could have been a bear after all."

"No bear. It was Sasquatch," said Grandma.

"Naturally, Dr. Stone is on its trail," Lucy continued. "It's so tremendously exciting!"

"Boy! It's fantastic! If it can be proven that it was Sasquatch then no one will *ever* doubt its existence!" Steve could hardly contain his excitement.

"I no doubt Sasquatch."

"We know, we know, Grandma. We believe you. But the others don't."

"They are fools," she said grandly, as she turned back to the stove.

"Dr. Stone's life dream might be coming true. It blows my mind!" Lucy trembled.

"What can I do to help?" Steve asked eagerly.

"I can't think of anything, really...I don't expect any messages for a few days. Dr. Stone will be on the trail following the footprints. But to be on the safe side, help me at the transmitter from noon to one every day, okay?"

"Sure!"

Breakfast over, Steve followed Lucy to the community center.

It was dark and it felt a little damp in the lab. Lucy flipped the switch on, and the fluorescent tubes under the ceiling blinked and hissed for a moment until the brilliant white light flooded the basement.

On the trestle table along the wall was a log book and a microscope covered with a large plastic bag.

Steve read Lucy's entries in the log. "Awesome! Let's cross our fingers that this time Dr. Stone will be successful in his search!"

"Let me show you the transmitter," Lucy said.

"No problem. I know how to operate it. It's no more complicated than a CB radio."

"Okay, then. Let me show you the latest negatives." Steve followed Lucy to the photo lab, feeling doubly happy at her nearness and the prospect of working with her.

The photo lab was a converted bathroom. The red light under the ceiling gave the small room an atmosphere of coziness. The wash basin was filled with photographic trays stacked to dry as if they were dishes in the kitchen sink. Under the low wooden beams Lucy had strung wires to which she attached strips of developed negatives, holding them like laundry with plastic clothespins of many colors.

Steve took down one of the strips. He turned on the high intensity lamp, and peered at the negatives against the light.

He saw masses of underbrush and a vague form of some creature seen from the back. It looked as if it was standing erect but it could easily have been a bear or even an old tree stump covered with moss that had turned dark with age. Even passionately wishing to believe that it *was* Sasquatch, Steve could not recognize the familiar outlines of a humanlike body in this not-too-clear negative. "Dr. Stone must come up with something better than these pictures," he said with disappointment. "These photos prove nothing."

The Chief and Joey arrived from Victoria in the afternoon of the following day.

The helicopter circled over the village, hovering as if suspended in the air. It landed gently over the meadow, its rotor blades sending blasts of wind that flattened the grass and tore the kerchiefs off women's hair. The villagers crowded on the periphery of the meadow aware of the deadly rotor blades, but as the revolutions stopped, they surrounded the machine.

The door opened and a pale, but smiling, Joey appeared in its frame. The towering figure of his father loomed behind him. Jim helped Joey to the ground.

The villagers rushed to greet him. The pilot saluted. Then he turned the ignition on. Hastily the villagers retreated as the rotor blades moved, slowly at first, then faster and faster. With a screeching noise the helicopter rose. It flew over the meadow in a broad arc, the pilot waving to the people below.

"Well, Tiger, back to bed," the Chief said, his hand on Joey's shoulder. "The doctor made me promise that for the next week you're to stay put in the house. No chasing around."

Joey made a face. "Okay...can the guys visit me?"

"Sure." They walked slowly up the hill toward their house, both wondering why Steve was not among the greeters.

Steve did not hear the arrival of the helicopter. He was helping Lucy to tidy up the lab since early morning, listening to her stories about Dr. Stone's research. She had been Dr. Stone's assistant for two years, and she worshipped him. "My boyfriend teases me that I'm in love with Dr. Stone," she laughed.

"Your boyfriend!" Steve felt blood rush to his face.

"Sure. I'm engaged to be married this fall," Lucy said not noticing Steve's discomfort. "Wanna come to my wedding?"

"Sure...no...I don't know..." he stammered

"I was just kidding. You'll be in school in California by then," she said.

The door to the lab was thrown open and Dan appeared in its frame. "Joey's back!" he yelled.

"Great!" Steve felt relieved. His conversation with Lucy had become painfully embarrassing to him.

"Let's go!" Lucy reached for the keys to the lab. They hurried to the Chief's house.

Joey was stretched out on the sofa in the living room.

"Hey, Joey, how're you feelin', man?" Steve greeted him.

"I'm okay. It hurts, but only when I laugh. You know, like the brave with an arrow in his back. Asked if it hurts, he said, 'only when I laugh.' You know the joke."

Steve nodded. "Yeah I know it. It's at least a million years old!" He was glad to see Joey in good spirits again.

"I hear you're a big shot now. The moment I got in, Grandma told me that you're working with the transmitter and that Uncle Herbie had seen Sasquatch!"

"Is very important," Cooing Dove said sternly. "Herbie discover Sasquatch—the whole world make him big hero."

The friends and relatives crowding the room laughed.

"Yeah, we all will become the most famous people in the world!" Jim said. "Thousands of tourists will flock to our humble village. We'll be rich and famous and drive only Cadillacs."

"And we'll build a high-rise Hilton Hotel," Sandy joined him. "I'll be the social director and drive a Porsche."

"And the community center basement will become a museum," Jim continued. "The Sasquatch Museum!"

"And Dr. Stone will be the curator!" Sandy was taken over by the joviality. "And Steve will be his assistant. He's one of the true believers in Sasquatch!"

"I proudly accept the appointment," Steve bowed with pretended pomposity. "Actually, I know a little about museum work. Our school has a special program at the Los Angeles Museum of Natural History."

"It must be cool to work in a museum," Joey said, a touch of envy in his voice.

"It is. Although it is a bit scary too when you realize that all this expensive equipment is there. Even in Dr. Stone's little lab, in that narrow room...what if something happens?"

"Like what?"

"Well, I don't know, the rats might chew the films, or an earthquake might collapse the building..."

"We have no earthquake," Grandma said. "For rats I give you a cat."

"He didn't mean it, Grandma," Joey said. "Nothing is gonna happen to Uncle Herbie's equipment. Steve just takes everything too seriously."

"He must take it seriously. It proves to people that Herbie find Sasquatch!" Grandmother said, her chin raised stubbornly.

Standing at the door, lost in the crowd, was Chickie. He had come to his uncle's house with the rest of the relatives. He took no part in their cheerful banter, but when Steve mentioned his apprehension about the safety of Dr. Stone's equipment, Chickie's eyes suddenly sparkled malevolently behind his thick lids. As he watched the visitors crowding around Joey, Chickie's mind began formulating an idea. He knew now how he could destroy his enemy, the White Belly kid who had humiliated him by winning the archery competition.

Unnoticed by anyone, Chickie slipped out of the house. He had to think over his plan and make preparations.

He felt elated. He knew now what had to be done!

X

Chickie's Revenge

    Chickie ran toward his house on the outskirts of the village, muttering under his breath, "I'm gonna fix you now, White Belly! You just wait and see!"
    He passed the community center noting that every transom window of the basement was locked. No matter. Nothing could stop him now from entering the lab once he made up his mind.
    He waited for the night to fall, hiding inside a Lincoln Continental left rusting on its axles at the back of his house. The once expensive car had been wrecked by his father years ago, and stripped of all its usable parts. Now only the frame remained, dented and twisted beyond repair. The upholstery, once a rich red leather, had been removed by Chickie's father to make belts and moccasins. The exposed stuffing slowly disintegrated, broken coils protruding out of the seats like upended corkscrews. Through the shattered windows, birds flew in and out finding the wreck a convenient place for nesting.
    Chickie, too, found the car convenient. No one ever bothered him there. If the village children came near his refuge, he pelted them with stones. It pleased him to watch the children scatter. He would swagger along the main street later, watching them scurry out of his way like a school of small fish at the approach of a shark. Even the village dogs gave him a wide berth whenever

they chanced to be in his path: his aim with a stone was accurate.

"I don't need nobody!" Chickie kept telling himself, spitting at his feet.

But underneath this shell of arrogance, Chickie craved acceptance. He would have loved to be like Joey, admired by everyone; or like Dan, chosen by the kids to be the captain of the archery and the baseball teams, but he was never proposed for the position. He was too mean to be emulated, too treacherous to be trusted. Even his excellence in archery never gave him the popularity that he craved. No kids trailed after him following a victory at a competition like they trailed after Dan and Joey. After his poor showing at the potlatch archery competition, the kids ignored him.

"I hate them *all*," Chickie thought.

Chickie stirred in the back seat of the demolished Lincoln. He crossed his arms behind his head and stared at the darkening sky through the jagged edges of the broken glass.

He had no specific plan. He only knew that he was "gonna fix Steve Bradley" by destroying Dr. Stone's laboratory.

Chickie hardly ever made plans. He always acted impulsively, never thinking of the consequences. When his hatreds took hold of him, Chickie was incapable of clear thinking. He would abandon himself to his passions, breaking windows or setting fire in a classroom if he happened to be angry with a teacher.

When caught—and he was always caught—he took his punishment stoically. But no punishment, no beating from his father, no appearance in a juvenile court, no ostracism by his peers, had ever made Chickie regret any of his violent outbursts or apologize for his actions.

Eventually, every broken window in the village, every missing hubcap was attributed to Chickie. He never denied such acts of vandalism whether he had committed them or not. He accepted the punishment in silence, revelling in his self-imposed role as "the meanest kid on the Northwest coast."

As Chickie listened to the gradual descent of silence, waiting for the night, he felt the wave of excitement rising in his chest. He was going to revenge himself for his defeat at the Nootka potlatch!

He watched the lights in the houses disappear one after another. Soon only the blinking neon sign at the service station was still visible, but Chickie knew that the service station was already closed. The sign remained lighted through the night and presented no obstacle to his plans.

He saw Lucy, accompanied by her huge dog, cross the road toward Sandy's house where she was staying.

A dog barked in the distance and was answered by a chorus of other village canines, all doing their duty.

Chickie waited. Finally, the village grew totally silent. Even the crickets became drowsy, chirping lazily now and then. Only the owls were wide awake hooting in the darkness, ready to swoop down on some small unsuspecting creature.

Chickie waited for another half hour, listening to the owls, savoring the forthcoming humiliation of Steve Bradley.

At last he decided it was time to move. He reached under the coils of the front seat and retrieved a crowbar. Probing still further, his hand came upon a can of spray paint. Chickie grinned. He surely could use it tonight. Still probing, he felt the streamlined form of a flashlight. He checked the batteries, carefully shading the light with his hand.

Everything was in order. Noiselessly, like a true scout, Chickie slipped out of his hiding place. Bending low, he ran from tree to tree, standing motionlessly for a few moments in the shadow of each one, keeping away from the houses, careful not to alert the ever vigilant watchdogs. Presently, he reached the back of the community center with its basement windows facing the silent meadow.

With one swing of the crowbar, Chickie smashed the first window. The glass splintered and fell on the ground with a clear, sharp sound.

Chickie froze, holding his breath. The tinkling sound reverberated in his ears as if it were a loud explosion. Any moment now there could be a wild cacophony of the dogs' barking, and the thunder of running feet as the whole village converges upon him.

But all was quiet. No one had heard the sound of the broken glass.

Slowly Chickie let the air out of his lungs. His heart was pounding but he felt no fear.

He removed the remaining jagged pieces of glass from the window frame, piling them along the wall. He had no need to cover his tracks. Everyone would know in the morning that it was *he* who was responsible for the destruction in the lab. And everyone would blame Steve Bradley. Were it not for Steve, Dr. Stone's lab would've been safe!

In his illogical and hate-filled mind Chickie did not think of the harm he would bring to Dr. Stone, nor the severe punishment he himself would receive for his vandalism.

Obdurately, he could only think of his enemy—Steve Bradley.

Chickie climbed through the transom window and easily slid down to the floor. Moonlight poured through the windows leaving oblong patterns on the floor, providing almost enough light, making his flashlight unnecessary.

"Cool!" he whispered.

The microscope caught his eye. He lifted it off its rubber pad and hurled it against the cement wall. The expensive, delicate instrument fell apart in dozens of pieces, its lenses smashed into tiny shards, its metal parts bent and useless.

Chickie felt excitement welling up in his chest, almost suffocating him in its intensity, making his blood run faster, his heart pound and his hands sweat. He began to whistle "Old MacDonald Had a Farm," as was his habit when he was agitated.

Next he disposed of a laptop computer. He dealt it a mortal blow by ramming his crowbar into the heart of the machine, twisting it ruthlessly as he pulled at the keys, tearing them out like so many broken teeth.

His excitement reached its zenith when he discovered the photo laboratory in the converted bathroom. He knew that Steve was helping Lucy with the development of the films. Whistling faster and louder, Chickie demolished two cameras and smashed the photo enlarger against the hard cement floor.

He felt as if he were floating on high waves. He felt exhilarated, freed of his earthly body, invincible in his power. Still whistling "Old MacDonald," he concentrated his fury on the transmitter. He lashed at the instrument crushing every component with the crowbar.

Chickie felt no curiosity about the intricate instruments he was destroying so ruthlessly. He was blind to anything but the image of Steve. It kept throbbing in his mind's eye like a strobe light.

The basement was a shambles. All around Chickie were strewn twisted parts of the expensive precision instruments. Broken glass crunched under his feet. With relish he ground it even more with the heels of his boots.

His glance fell on the daily log. Delighted in finding yet another means of desecration, Chickie turned the pages of the log spraying paint on each of them. Soon the whole journal was covered with the sticky red paint, dripping like blood. Chickie slammed the covers shut satisfied that nothing could be deciphered from its pages.

"What else?" he asked himself looking around, still seeking, not yet sated.

He went back to the photo lab and emptied all the chemicals into the toilet. For good measure, he smashed the bottles as well. Then he ran the crowbar along the walls leaving a deep wavy scar on the newly painted surface.

Finally, everything was totally destroyed. "Well done!" Chickie told himself. One more thing remained to be done—leaving his signature.

Using the remnants of red paint, Chickie sprayed a broad

"Z" on the walls of the basement laboratory, the mark of Zorro, the sign he had expropriated as his personal signature from an old Western movie.

He was satisfied. The lab was destroyed. Steve Bradley would be crushed. He would be blamed for the lab's destruction. Steve Bradley, the big shot from California who should've been watching the lab, as he was told to, instead of acting like a movie star at the potlatch, parading around with the best bow and arrow set. "Steve Bradley, White Belly, you're a cooked goose," he thought.

XI

Honor Bound

Cooing Dove discovered the ravaged lab early next morning. Crossing the meadow on her way to the Chief's house, she passed the community center, noticing a broken window. Curious as always, she peered inside.

A picture of devastation spread before her. Gathering her skirts, she rushed to the Chief's house, screaming as she ran.

"Herbie's things smashed!" She burst into the house. "Go quick. All smashed!"

"Calm down, Mother, tell me what happened?" The Chief stood in front of the mirror, shaving.

"Dunno. I saw window broken in the basement. Everything smashed to pieces!"

The Chief pulled a sweater on.

"What happened?" Joey's sleepy voice came from the sofa in the living room where he was sleeping during his recuperation.

"I don't know yet. Grandma says that Dr. Stone's lab has been vandalized."

Steve, awakened by the commotion, came into the living room. "What happened?" he asked, yawning and rubbing his eyes.

"Grandma says that Dr. Stone's lab has been wrecked."

"No!" Steve rushed to the door, his sleepiness gone.

"Get dressed first," Grandma grabbed him by his pajama

tail.

Hurriedly he dressed, dashing out the back door, his sneakers untied, the laces trailing over the dewy grass.

"You stay in bed!" Cooing Dove said seeing that Joey was about to follow him.

"Yes, Joey, don't add to our problems. Remember, you're still postoperative. Stay put," his father said.

"Oh, okay," he sighed, falling back on his pillows. "But hurry back and tell me what happened."

Cooing Dove and the Chief ran after Steve.

They froze at the entrance to the lab, bewildered. The Chief stepped inside, broken glass crunching loudly under his feet.

"Someone, get Lucy. Quick!" the Chief ordered.

"Everything, *everything* is smashed!" Steve cried out. "Who would ever do such a thing!"

"Must be someone hate Herbie Stone," Cooing Dove said.

"Nonsense! Our people love Herbie Stone. He's been our tribe's best friend for many years," the Chief said. "No one would want to destroy his property. No, it must be someone from the outside...some tramp. None of our people would ever do such a thing!" The Chief walked around the overturned tables, his eyes flashing and his jaws set in a grim expression of anger.

"I bet I know who did it," Dan said softly. He and the others had been awakened by Grandma's loud cries and now they crowded at the door.

"What are you saying? Speak up!" the Chief turned to him sharply.

"I think it's Chickie...I mean, Billy Wallace."

"Billy Wallace? My own nephew? Why do you say that?"

"He always leaves the sign of Z. It's his signature. All the kids know it. You know, the mark of Zorro, from the old Western..."

Everyone looked at the huge red Z's on the walls.

"Get Billy here! On the double!" the Chief growled.

Dan took off for Chickie's house.

Lucy was out of breath when she rushed onto the lab. Stunned, she shuffled aimlessly among the wrecked articles, picking up a twisted wire from the transmitter and a smashed lens from the microscope. She picked up the journal but its pages were firmly glued together with the dried red paint. Disheartened, she dropped it. Tears welled up in her eyes as she scanned the carnage.

The villagers, arriving by the score, peered with curiosity through the windows, aghast at the senseless destruction.

"Nothing spared," Cooing Dove informed them. "Herbie Stone be very sad."

"Sad!" Steve thought. "He'll be devastated! Thousands of dollars worth of equipment—all destroyed!"

He heard the noise at the door. The crowd parted to let Chickie enter.

He was walking between Jim and Dan, as if he were their prisoner, an arrogant smile on his face.

"He bragged that he did it, Uncle," Jim said.

"Did you do it?" the Chief demanded, with barely controlled fury.

"I sure did!" Chickie laughed. The chaos in the lab looked even better to him in the daylight.

"Why?" Steve yelled. His voice cracked and the question came out high and squeaky, as if he were a small boy.

"Why?" repeated the Chief.

"Because I hate his guts!" Chickie sputtered, the smile gone from his face as it turned ugly with hatred. Saliva had begun to gather and bubble at the corners of his mouth.

"What had Dr. Stone ever done to you that you hate him so much?" Lucy burst out.

"Who's talking about Dr. Stone? He ain't done me nothing. I don't hate him. I hate Steve White Belly Bradley!" Chickie spat out vehemently.

A hush spread over the crowd, broken finally by Grandma: "Billy Wallace, you crazy!"

"I ain't crazy! I always hated his guts. Ever since he arrived here, I hated him! Yeah, I wrecked the lab because *he* worked here," Chickie shouted his face beading with sweat. "Yeah, I wanted to see the expression on his stupid face when he saw the lab!" The sweat poured down his face.

"I'll kill you!" Steve rushed at him, but Dan quickly pinned his arms down. The Chief put a restraining arm on Steve's shoulder.

"He's crazy. Bad spirits stole his soul," Grandma said. "Call the shaman."

"Get his father. We must notify the juvenile authorities," the Chief said.

Jim tied Chickie's hands behind his back. Chickie did not struggle. He laughed, watching Steve's despair. "I fixed ya, White Belly, didn't I?"

"You're a disgrace to our tribe! This time you'll stay in a reform facility for as long as the judge will sentence you. The tribe is not going to bail you out any more, Billy Wallace. Even though you are my own nephew, I am not going to plead on your behalf. You'll go to jail and stay there," the Chief said.

"Big deal!" Chickie snorted. "It won't be the first time."

The Chief turned his back on Chickie.

A pall of gloom spread over the village. With the benefit of hindsight, everyone now recalled some episode of Chickie's meanness which made the events of the previous night seem predictable. People told one another that they had *always* suspected that Billy Wallace was crazy. Many had admitted that they were always afraid of him. Everyone felt relieved when Chickie's father and Jim took him away to be booked at the Juvenile Detention Center.

Steve and Lucy began to clean the ransacked laboratory.

"I should have slept at the lab at night. It's all my fault," Lucy kept repeating. "I should've been a better keeper."

"Don't be silly. How were you to know that Chickie would become crazy and wreck it?"

"I should have never left the lab unattended," she lamented.

"We must let Dr. Stone know about the destruction of his lab," Steve said.

"How?"

"I'll go and look for him."

"How are you going to find him?"

"I have a map. It was because of me that his lab was destroyed. I can't let him come back and find his equipment all gone. I must prepare him for it."

"You don't know anything about the forest," Lucy said. "It's like nothing you've seen before. It's a rain forest, like a jungle. No trails. Nothing."

"Do *you* know it?" Steve challenged.

"No, I don't. Even though I grew up around it."

"I said, I have a map."

"Don't be silly. Dr. Stone more likely is in some unmapped territory by now."

"I'll find him. I'll ask Dan to go with me."

"Where do you want to go?" Dan asked from the doorway. "I came to help with the clean up," he explained to Lucy.

"How about being my guide?" Steve said.

"Where do you want to go?"

"I must find Dr. Stone."

Dan was taken by surprise. "You're kidding!"

"No, I'm not. How about it?"

Dan pondered the proposition. He knew that the Chief would forbid such an expedition.

"I'll go alone if you don't come with me," Steve threatened. "Don't you understand? I must go! It's my duty to go. It's my honor. You, a Native American, better than anyone, should understand when I say 'my honor.'"

"Okay, I'll go," Dan decided. "But we shouldn't tell anyone. They'll forbid our going. When shall we start?

"Now."

"Don't be silly..." Lucy began but Steve raised his hand.

"Please. I've made up my mind. I *must* go. Just don't tell anyone what we're going to do. At least, for a few hours, don't tell anyone, okay?"

"But it's very dangerous.... The Chief would never forgive me for not telling him. What should I do?" she was close to tears.

"Don't do anything, just don't say anything for a few hours to give us a head start. No use procrastinating, let's get going. We must prepare our gear," Dan said. He took over the arrangements.

"You're a real pal!" Steve clasped his shoulder. "I truly appreciate it."

"Aw, quit that," Dan shrugged off his gratitude with an embarrassed laugh. "What are friends for? Git goin', man!"

A couple of hours later the boys, bearing backpacks with their sleeping gear, cooking utensils and food, were ready to leave. Dan also carried a bow and a quiver full of slender arrows over his shoulder. "Just in case we have to defend ourselves against a big, bad bear," he joked.

"We might meet Sasquatch instead of a bear," Steve grinned as he reloaded his camera.

They confided their plan to Joey, swearing him to secrecy. They hoped that no one had seen them leave, as they hurried across the meadow, but Cooing Dove spotted them from the kitchen window. She knew at once where they were headed.

"Steve and Dan go to find Herbie, no?" she demanded bringing Joey a bowl of soup.

There was no use lying to her. "Yeah," he admitted. "They went to tell him the awful news. Steve still thinks that it was his fault."

"Not his fault. Billy bad guy. But is good the boys go to Herbie. Behave like real braves!"

"You're not angry that they went? Even though it's dangerous?"

"No, I'm not angry," she smiled her nearly toothless smile. "Everything dangerous. Cooking soup dangerous. Life dangerous.

The boys no fools. They watch for danger. They take no stupid chances."

"Danny is a good scout."

"Sure. He has a good teacher—his grandpa Big Beaver Tooth."

"But don't tell dad, okay?"

"Okay." She ruffled his hair affectionately.

"Where's Steve?" the Chief asked seeing that only two plates were set on the table. Grandma rattled the dishes loudly, pretending not to hear. "Is he still brooding over the lab? Steve!" he called as he went toward the boys' room.

Joey saw no way of avoiding the truth. "He has gone to find Uncle Herbie."

"What?" the Chief turned around sharply. "And you didn't stop him?" he raised his voice alarmingly. "Don't you know that the safety of Steven Bradley is my personal responsibility? He's an inexperienced city boy who doesn't know a thing about survival in the wilderness!?"

"But Beaver Tooth does," Grandma said turning from the stove. "Danny Beaver Tooth good scout. You say yourself Danny is a good scout. Don't blame Joey."

"I am sorry," the Chief said. "I don't suppose you could have done anything to stop them."

"I didn't even try to," Joey said defiantly. "Steve is right, he must be the one to break the bad news to Dr. Stone. Chickie had no bone to pick with Dr. Stone. But he hated Steve ever since Steve won the archery competition. Even before. If it weren't for Steve working at the lab, Chickie wouldn't have vandalized it."

"Yeah, Billy no hate Herbie. He hates only Steve," Cooing Dove came to the help of her favorite grandson.

The Chief pondered their arguments. "Steve shouldn't have gone to the woods without my permission. He could have gone there with a scout and still be the one to notify Dr. Stone. I was going to send a scout to his camp tomorrow..."

"Don't worry, Steve be okay. Danny good scout. They are smart boys!" Grandma said.

"Yes, Dad, don't worry. Steve has a map and a compass. And Danny knows these woods and mountains."

The Chief was silent, thinking that perhaps these few days in the primeval wilderness would be a great experience for Steve, one he would cherish all his life. Finally, he smiled at Joey. "Well, I suppose we owe it to Steve. If he feels that it is the honorable thing to do, we'll let him do it. Let's hope that Dan is as good a scout as we hope he is and that Steve's map is accurate. But just the same, to be sure, I'll send an adult scout on their trail. Just in case.... He'll stay out of sight, but in case of need, he'll be there, ready to help."

"Dad, you're the greatest!"

"Yeah, you pretty smart cookie," Grandma said, beaming at her two men.

XII

In Search of the Expedition

The boys walked steadily for over three hours. Dan wanted to cover as much distance as possible, in case the Chief sent a party in their pursuit.

Steve had never been deep inside the rain forest before. He had explored its approaches with Joey, but their promise to Dr. Stone, to stay away from it so as not to scare Sasquatch off, prevented him from going deeper into the forest.

He feasted his eyes now on the wild beauty surrounding him. Huge trees rose straight toward the sky, their thick trunks entwined with garlands of creeping vines. A mass of giant ferns covered the forest floor, vying with prickly berry bushes for every available inch of ground. Delicate flowers on long thin stems clustered here and there, providing a welcome splash of color against the dark greenery.

The path felt soft and damp under his feet, and it was often slippery. It was quite cold out. The sun never reached the path through the thick vegetation, touching only the tops of the trees with its warm rays. Steve was glad that he had on a heavy Alaskan sweater that he had exchanged with a Haida archer for his Beverly Hills High sweat shirt.

The path began to climb. "We should reach the firebreak soon," Dan said over his shoulder. "We'll rest there. But I want to strike camp for the night at the waterfall in Eagle Canyon. It's only about five miles further."

"Okay with me." Steve was relieved that they were so close to their campsite, unaware that these five miles would be so difficult to traverse.

They reached the firebreak. Dan loosened the backpack over his shoulders.

"Let's chew a few strips of beef jerky. We'll fix something warm when we reach the campsite. There is a little shack and plenty of dry firewood to build a fire."

"Firewood! How do you know?"

"It's our Native custom. Whenever we use a shelter, we always leave enough firewood for the next person."

They propped their backs against a jagged tree stump which sprouted new saplings from its sides. Below them spread the lush green valley, and beyond it, the dark blue Pacific.

"It is so beautiful!" Steve said looking at the panorama before him. "It's so quiet here, as if only you and I are left on this planet. I know the woods are full of animals, yet I haven't seen any."

"You must learn to move without making so much noise."

"I don't make any more noise than you do," Steve protested.

Dan smiled. "Sure. It's just the things that rattle in your pockets."

Steve stuck his hands into his pockets and extracted a Swiss army knife, a handful of small coins and a set of house keys. "Gee, I didn't know I had so much junk in my pockets!"

"But the animals knew. They heard you coming and went into hiding. We have a Native rule: never allow anything to clank."

"Boy, at the end of this journey I ought to be a seasoned scout myself," Steve laughed, redistributing the coins and the keys among the outer pockets of his backpack, but keeping the knife within easy reach.

"Yeah, you might as well be. Our people have adopted

you already after your victory at archery."

"Honestly?" Steve blushed with pleasure.

"Yep. Anyway, it's time to get going. We must reach the shack before darkness."

Steve wished that they could stay at the clearing a little longer, perhaps even take a nap, but Dan was already adjusting his backpack.

"Okay, let's go," Steve said, rising, and threading his arms through the straps of his pack.

They followed the firebreak which seemed to bisect the mountain like a broad scar. Below them spread the dark Eagle Canyon. Somewhere there, among its jungle of trees, was Dr. Stone's camp. Somewhere there, could be the hideout of Sasquatch, Steve thought.

"When we get down into Eagle Canyon we'll be in virgin territory. No more paths. We'll have to make our markings for the return trip," Dan said pointing to the green mass of trees below.

They began their descent. Dan walked lightly, his feet in soft moccasins barely touching the ground. Steve felt, in contrast, that he was trudging like a clumsy mastodon in army boots. Going down the steep mountain was hard on his leg muscles; his pack weighed heavily on his back, pulling him down.

"Tough, eh?" Dan said over his shoulder. "Let's revert to another native custom. It takes a bit longer but it's much easier on the legs." He began to walk in a zigzag, following the contour of the mountain, moving a few paces in one direction, then reversing himself, descending all the time.

"We've learned it from the animals," Dan said. "It's seldom that you see the animals going in a straight line, up or down. They always follow the contour of the mountain. Of course, mountain goats are the exception. And speaking of goats, look over there. Those are mountains goats," he pointed.

Steve squinted against the sun. On the other side of Eagle Canyon he could see several white specks moving among the craggy boulders high above the tree line.

The boys rested, watching the goats climb almost vertically along the bare face of the mountain. The animals moved with amazing speed, noticeable even from this great distance.

"Something must have scared them," Dan said. "Must be a cougar or a bear."

"Or Dr. Stone and his crew," Steve joked.

"It could also be Sasquatch."

The familiar feeling of anticipation expanded in Steve's chest. It felt almost painful, this expectation of something truly significant about to happen.

"Do you think we might meet with Sasquatch?" he said, his voice sounding shaky.

"Once we're in Eagle Canyon, anything is possible. It's one of the least explored canyons in the world. But let's get going. I want to reach the shack before dark."

The boys resumed their zigzag climb down the mountain.

The canyon was deep and narrow. It was huge, its twisting miles covered with a thick evergreen forest of hemlocks and cedars. Although the sky over their heads was bright and the sun illuminated the top of the mountains, it was dark in the canyon. A cool breeze smelling of swamps and decay rose from its bottom.

"It will be cold tonight," Dan said. "But I prefer it to a balmy night. When it's cold, the mosquitoes don't bug you."

Steve was growing tired. He wanted to rest, but Dan relentlessly plodded ahead.

There was the sound of rushing waters and a broad stretch of churning white waters spread below them.

The stream was wide, more like a river, strewn with huge rocks. The water surged between the rocks, white and foaming, roaring as it rushed along the stiff banks.

"Right on the dot!" Dan said. "I haven't been here for a couple of years but I found it all right!"

With renewed energy, the boys ran down the remaining few yards to the water.

Dan flattened himself on the rock jutting out of the stream.

He lowered his face into the water, sucking it into his mouth in thirsty gulps. Steve followed his example. At once his nose became full of water. He burst out coughing, his sinuses smarting.

"Hell, don't tell me there is some secret Native way to drinking water out of the stream!"

Dan laughed. "Just hold your breath. You surely wouldn't inhale while swimming under water, would you?"

Steve tried again. This time he allowed the water to filter into his mouth. He swallowed, still holding his breath.

"Now you got it. Doesn't it taste terrific? Our shamans prescribe mountain spring water as their secret medicine!"

They rested on the rock, no longer in a hurry. There was still light, although the sun was obscured by the mountain, as it cast a dark shadow across the canyon.

"You'd never believe that it's still August," Steve said. "It's cold!"

"Okay," Dan said sliding down the rock. "Let's go. We must cross the stream. I'll go ahead." Probing ahead with a long stick Dan made his way across the stream, jumping from rock to rock, pausing on a boulder in the middle of the stream.

"It's your turn now. Use a stick to balance and to probe ahead. Don't look into the water. Go!"

Cautiously, Steve stepped off the boulder trying not to look at the rapidly moving waters inches away from his feet. Balancing a heavy pack, his camera swinging over his chest, Steve's progress was painstakingly slow.

"Boy, it's tough!" He was breathing hard.

"Yeah. Worst of all, there's no hold for your hands, should you slip. When we travel with small kids, we chop down a couple of trees and drop them across the water like a bridge. Perhaps we should have done it," Dan said, watching Steve, ready to jump into the swirling waters, should Steve need help.

The muscles of Steve's legs trembled from exhaustion. Every new step was becoming more difficult, and Steve was landing on the rocks with less precision.

At last he reached the boulder in the middle of the stream. Dan stretched out his hand for Steve to grab. "You couldn't have done it better if you were a Native!" he said, pulling him up.

"Gee, thanks! I never thought it would be so hard. I have been jumping on the rocks ever since I was a Cub Scout."

"But never in Eagle Canyon."

"You said it. Never in Eagle Canyon!"

"Okay, let's keep going. I'm getting very hungry," Dan said. They forded the remaining shallow part of the stream. Once on the ground again, they followed the riverbed.

Fallen trees blocked their way, forcing the boys to climb over the slippery, moss-covered trunks, their feet often crushing into the soft bark. They heard the sound of cascading water, gradually increasing in volume as they approached it. Suddenly, as the riverbed curved, magnificent falls rose before them.

"Wow!" Steve shouted, but Dan could not hear him. The roar of the water was deafening. The boys paused at the foot of the falls, staring at the water as it rushed down from the top of a high cliff over several levels, splashing into a fine spray over the crags, churning at the bottom, as if boiling.

"Fantastic!" Steve yelled, poking Dan in his back. Dan nodded, grinning.

The air was filled with mist and the trees around the falls glistened with droplets of water. Dan pointed to the woods.

"Let's go! It's not far now," he shouted. He made a sharp turn away from the falls and plunged into the forest.

It was dark now but Dan moved with the assurance of one who knew where as was going.

Then he stopped. "Here we are. Our 'Holiday Inn.' We made it!"

Steve saw a tiny clearing. Against one of the trees there was an A-shaped shack made of saplings. The shack had a forlorn look as if no one had visited it for years but there was a supply of dry twigs piled neatly inside.

"Am I hungry!" Dan freed himself from his backpack.

"We're going to eat an enormous supper!" They spread their sleeping bags inside the shack and unpacked their food supply—beef jerky and dried fish.

Steve thought of the dehydrated gourmet dinners, the kind of meals one could buy in the specialty stores in Beverly Hills. None of that fancy stuff here. Dan had assembled the food according to Native customs.

"It's okay with me..." Steve thought. "If the Natives can survive on dried fish and beef jerky, so can I." He stretched in front of the fire which Dan had already started, thinking what a terrific adventure his summer vacation was proving to be, for a moment forgetting the reason that he was in the forest.

They slept soundly, undisturbed by the roar of the falls. In the morning after breakfast of more dried fish and jerky, they replenished the supply of wood in the shack, and doused the dying campfire with water, making sure that no spark was left to be fanned by the wind into a roaring inferno.

"We'll follow the stream," Dan said. "Dr. Stone's camp should be somewhere around here." He pointed to a junction of two small rivers on Steve's map.

Once again they were on the move. They stayed close to the twisting course of the stream, chopping their path through the dense underbrush. Invisible sticky cobwebs, suspended from the branches, brushed against their faces, creating an unpleasant, gluey sensation.

Steve was amazed at the variety of deformed stumps and fallen trees, covered with a thick growth of moss, which obscured their original forms. Some looked like huge bears rising up on their hind legs, front paws trailing garlands of ivy. In the semi-darkness of the forest, one could have easily mistaken such a furry stump for a bear—or Sasquatch.

Other formations looked like the crumbled, crenellated walls of ancient fortifications, green with age. These were giant cedar trees toppled by storms or disease, rotting on the damp forest floor, housing hundreds of animals, including parasites, rodents and snakes. Through the years the decaying wood had become fragile, deteriorating into a mass of fetid pulp.

The boys avoided the fallen trees. Their feet could have easily become entrapped in hidden holes full of poisonous splinters. They made detours around the trees climbing over the rotting trunks only when there was no other way.

Furry, gray rabbits scurried across their path. Curious about the boys, the animals allowed them to approach quite closely. Then, taking flight, the rabbits would pop up again a few feet away like so many jacks-in-the-box. Noses twitching, their front paws hanging down like the arms of a tired prizefighter, they watched the boys pass.

Flocks of birds, rising noisily as one, circled over their heads, settling down again as soon as the boys moved.

In the afternoon, the sky filled with dark clouds. Relentless, gentle rain descended upon the forest. The smell of decaying vegetation rose to the point of suffocation.

Drenched to the skin, the boys hurried to reach another shelter. It was less commodious and the rain lashed inside it through a hole in the roof.

Dan took off his backpack. "It will be a problem building a fire." He looked around for a pile of wood. It was there, neatly covered with old cedar branches.

"Boy, am I glad that you people take care of one another!" Steve said.

Dan piled the kindling into a small pyramid over a hollow in the wet ground. He struck a match and the dry wood caught fire at once. "Quick! Give me the bigger pieces! We must protect the fire from the rain." Steve hurriedly passed him several larger pieces of wood.

They built another pyramid over the twigs, protecting the flame inside. Soon the fire was blazing, too strong to be extinguished by the drizzle, the smoke and sparks escaping through the hole in the roof.

They took off their wet clothing and shoes, and hung them over the sticks to dry before the fire, their bare backs pressed against the sides of the shack.

"We may be stuck here for a while. It's practically impossible to move through the forest during the rain," Dan said, chewing on a tough piece of beef jerky. "Let's go to sleep." He slid into his sleeping bag.

Steve could not sleep. Every muscle in his body ached and he twisted in his sleeping bag trying to find a comfortable position. For the past two days of trudging through the jungles of Eagle Canyon Steve had not thought of Chickie. Now, the enormity of Chickie's crime rose from the bottom of his consciousness in all its ugliness. Steve could almost see the despair on Dr. Stone's face when he would bring him the shattering news about the wanton destruction. He could feel Dr. Stone's pain, and dreaded his own role as the bearer of bad news. The feeling of guilt, for being the cause of the destruction, kept gnawing at him like a toothache.

"For Pete's sake, stop sighing and tossing. It was *not* your fault that crazy Chickie wrecked the lab. Go to sleep," Dan murmured, from the depths of his sleeping bag as if reading Steve's thoughts.

Steve smiled. "Thanks." He closed his eyes, and almost at once fell asleep.

XIII

Dr. Stone's Discovery

Knowing nothing about the events which led to the destruction of his lab, Dr. Stone and his students followed a barely visible trail left by some large creature. They walked slowly along the turbulent river, their eyes lowered, concentrating on the ground under their feet, searching for the footprints. They stopped often to consult with one another; what seemed to be a footprint often was a mere indentation in the damp sand.

Ever since Dr. Stone announced that they had reached the heart of Sasquatch territory, Dave and Mike had felt the tingling sensation of anticipation, mixed with fear of the unknown. For two days now, they followed the river, looking at hundreds of depressions in the sand, none of them resembling human footprints. The young men were growing cynical, but Dr. Stone methodically photographed every depression, making notations in his journal. "We're getting there," he kept saying to his young companions.

They did not share Dr. Stone's obsession with Sasquatch. They enjoyed his stories about it but, like many others, considered them to be nothing but tall tales. To them, Dr. Stone was a lovable eccentric, whom they admired, but made fun of behind his back. The expedition itself, was a lark, a great way of spending a vacation, better than any summer job they have ever had. They enjoyed the expedition as if it were another backpacking trip. "It sure beats being a short-order cook at McDonalds," they joked among themselves.

Suddenly, Dr. Stone stopped. "There!" he pointed dramatically to the ground. "There's Sasquatch's footprint! Clear, as if he were here only a few hours ago!" He knelt, closely examining a large, five-toed print, clearly visible in the damp, hard sand.

"It's definitely *not* a bear's footprint," he said. "I have seen many bear footprints during the years of my search; this one is different. There is no doubt. It is a humanoid print!" Dr. Stone's heart was pounding wildly.

"There's another footprint," Mike pointed further ahead. "This one looks like a left foot."

Dr. Stone exhaled noisily. He was not aware that he was holding his breath. "Measure the distance between them. Dave, look farther ahead, there ought to be others. I think we came upon the most recent trail."

Mike measured the distance between the two imprints. "Three feet, five inches..."

"Right. It makes one footstep for Sasquatch." Dr. Stone's mouth felt dry with excitement as he focused his camera for a closeup of the first footprint.

"Here are three more," Dave shouted. "All about three feet apart, left, right and left.... Like footsteps."

"They *are* footsteps. Gentleman, we have found Sasquatch!" Dr. Stone proclaimed solemnly.

"Yahoo!" Mike and Dave yelled, jumping up and down like excited chimpanzees.

"Sh-sh-sh! Don't make such noise. You may scare him off. We must be very discreet now. Sasquatch is a shy creature." Dr. Stone bent over the footprints again. "He can't be too far away," he muttered to himself. "The prints are superb. No one would ever mistake them for a bear's." He knelt over each imprint measuring it, then wrote his finding in a journal. He was calm now. Reloading his camera, he photographed the imprints from several angles, a measuring tape spread on the ground next to each of

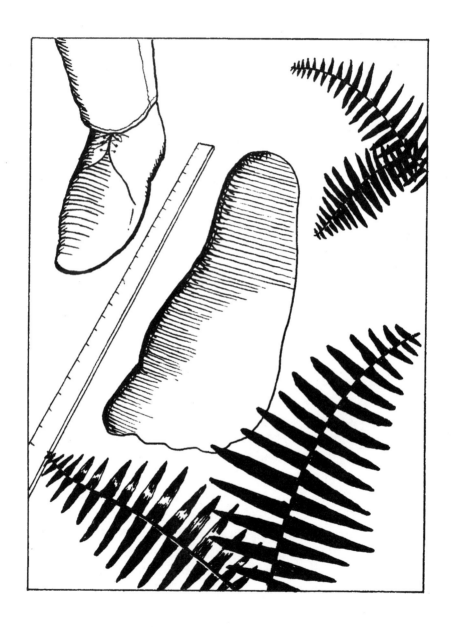

them. All the imprints were over eighteen inches long and almost eight and a half inches wide.

"This proves that the footprints belong to the same creature," Dr. Stone explained to his students. "Now, get to work. We must make plaster casts. Start mixing plaster, while I take more pictures of the general terrain."

The young men unpacked the gear containing plaster of Paris. They began mixing it, watching Dr. Stone covertly.

"Do you believe that we *really* found Sasquatch?" Dave asked in a low voice.

"So far, we found only a bunch of giant footprints. It's still a question: whose prints are they?"

"They are *his*," Dr. Stone said.

Mike and Dave exchanged glances. "The old man seems to have a perfect hearing," Dave thought.

The plaster was ready. "Should we clean the imprints before we pour plaster?" Dave asked.

"No, no, no!" Dr. Stone rushed back, his camera dangling over his chest. "No, no, I want the casts to show the footprints exactly as they are. We'll estimate the weight and height of the creature by the depth of the footprints and the condition of twigs and grass which he crushed under his foot. Okay, let's pour!"

The young men carefully poured plaster into all five imprints as Dr. Stone watched anxiously.

"Well done! Now, we must protect the casts," he said, satisfied that the plaster filled each imprint neatly. Brilliantly white against the dark, moist sand, they looked indeed like the footprints of a giant.

"We'll cover them with plastic and then with hemlock branches. Make sure though, that nothing touches the casts. By tomorrow, they ought to be hardened enough to be lifted," Dr. Stone said.

Mike and Dave cut several thick hemlock branches, while Dr. Stone spread sheets of plastic over the casts, securing the edges of plastic with river stones. Mike created little A-shaped covers over them using the hemlock, while Dave piled more stones around them.

"Perfect!" Dr. Stone said. "Now, since we have to stay here until the plaster dries, let's set our camp. Actually, it's a perfect place for a base camp. We have a river, where we can fish, we have a sunny clearing, which is rather rare in these dark, wet woods, and we have plenty of berries to gorge ourselves on...that's why we have had good luck in finding Sasquatch's footprints. This little meadow must be his favorite place. He visits it often."

"How do you know that it's his favorite place?" Mike challenged.

"Elementary, my dear Watson," Dr. Stone said. "Look closely at those bushes. Do you see the tufts of hair hanging on the thorns? They likely belong to Sasquatch. This is further proof that he was here."

Mike and Dave stared at the reddish brown tufts of coarse hair stuck to the thorny bush of red berries. "It looks like my dog's hair," Mike said, viewing the hair skeptically.

Dr. Stone was not offended. "It could be a bear's hair, of course," he said. "But until we examine it under the microscope, we won't know for sure, will we? I, personally, am positive that it's Sasquatch's hair. So, be a nice boy, gather a few samples from different branches."

Mike proceeded to pluck the tufts of hair off the branches. He placed them in individual small plastic bags, sealing each with tape and writing the date of collection on each bag.

Dave, meanwhile, finished unpacking their gear. He spread all the parts for their tents, cooking utensils and packages of dehydrated food on the ground. "Hey, Mike, help me string the rope," he called.

"I'll help you," Dr. Stone said. "Let him finish gathering the hair samples."

Dave climbed on a tree with a coil of rope. He tied one end to a stout branch and dropped the other end to Dr. Stone below.

"You'd better climb down, and up again on another tree," Dr. Stone called. "My tree-climbing days are over."

"Will do." Dave climbed up another tree, a few yards away, and tied the other end of the rope to a branch. "No bear—or Sasquatch—can reach that high," he said hanging the plastic bags containing their food on the line.

"Don't be so sure. Sasquatches are very curious creatures. It had been reported that they regularly raided woodcutting camps around here. As a matter of fact, I want to test Sasquatch's curiosity myself. Not with food, though."

"How?"

"I brought a cassette of Chopin's nocturnes along. I would like to play them tonight. Perhaps it will entice our friend to pay us a visit."

"He would probably prefer Bruce Springsteen or the Rolling Stones. They're more hip," Dave laughed, refusing to believe that Dr. Stone was serious.

"If you happen to have it, I am willing to try it, too, to enlarge my experiment. Although, personally, I believe that more melodious sounds will appeal to him more. I interviewed an old prospector one time, who claimed that he had seen Sasquatch, and had played the harmonica for him."

"No kidding?"

"Yes. The old man told me that as long as he played some cowboy songs in a minor key, Sasquatch stood quietly, listening. But as soon as he began playing "Turkey in the Straw," stomping his feet rhythmically, Sasquatch ran away. So, you see, the creature obviously did not like the lively tune, but enjoyed the slow, plaintive melodies."

"That Sasquatch had the soul of a poet," Mike said sarcastically.

"Don't laugh. There has been a lot of research done about the reaction of various animals to music. They all seem to prefer less raucous music. Perhaps, Chopin will be just the ticket."

Dave and Mike laughed, as they went about setting the

camp. "He sure is some character!"

They set up four small tents, one for each of them, and one for their equipment—radio transmitter, cameras, video camera, tins with unexposed films, and other small items, such as plastic bags for storing different samples. The camp was ready.

The sun was setting behind the mountain on the other side of the river, its rays gilding the massive cliffs above the tree line. Cool, damp air, smelling of decay, drifted toward the camp out of the river canyon.

Dr. Stone placed the cassette into his portable player. The sounds of Chopin's delicate music filled the air. Dr. Stone adjusted the volume and retreated into his tent. He was tired. He stretched out on his air mattress, his arms folded behind his head. He listened to the rippling piano music, as he thought over every detail of the past two days, ever since they came upon the trail. "Now, all I have to do is to find and photograph Sasquatch!" he thought. "I feel it in my bones, he's here, he's near—somewhere—hiding in these mountains!"

Suddenly, he stiffened. Like a hunting dog, he sniffed at the air, turning his head this way and that. There was an unmistakable sulfurous odor in the air, as if one were near a decaying marshland.

"Sasquatch!" Dr. Stone jumped to his feet, his tiredness gone, and grabbed for his boots. "Steady, steady," he told himself. "Don't panic. You have been waiting for this moment all your life. Don't scare the creature." Moving like in a slow-motion dance, he came out of the tent. He froze at its entrance, peering intensely at the thicket of woods at the edge of the camp. The odor was very strong now. Dr. Stone removed the cover from his camera, ready to shoot.

There was a sound of crushing branches. A large hairy creature came momentarily into view, only to disappear again, swallowed by the thick underbrush.

"Sasquatch!" Dr. Stone shouted, running after the creature. Dave and Mike jumped out of their tents, half-dressed.

"Get your cameras and video!" Dr. Stone disappeared among the trees.

The young men hastily dressed and, armed with their cameras, rushed after Dr. Stone.

They presently caught up with him as he stood panting.

"Are you all right?" Dave asked in alarm.

"Yes, yes.... I'm just out of breath. Let's go. We still have a couple of hours of light left. We're hot on his trail, let's not lose him."

The trail left by the retreating beast was clearly marked by broken branches, and it led straight up the mountain.

"There he is!" Mike suddenly yelled, pointing up.

"Take pictures! Start the video!" Dr. Stone shouted, running again in pursuit.

Halfway up the mountain a large, hairy beast was moving steadily through the trees. It was hard to see whether it moved on all fours or upright; it was intermittently hidden by vegetation, coming into full view only now and then.

"Don't stop the camera. Keep shooting," Dr. Stone shouted, climbing up the mountain after Sasquatch.

Mike could occasionally see the creature through his view finder. He ran, trying to keep the camera steady. "The film will be lousy," he thought, as he stumbled and fell.

Dr. Stone and Dave rushed ahead, snapping their cameras as they glimpsed Sasquatch through the trees. They wished he would slow down and turn toward them, but the creature ploughed ahead as if it knew no exhaustion.

Mike, running with the video camera behind them, shouted, "He's just above the tree line! There's a cave there...I see it clearly!"

"Keep shooting!" Dr. Stone yelled. "Thank God for the long northern days," he thought. "We still have at least another hour of light."

"He entered the cave! I don't see him anymore."

"A cave! Dear God, we've got him!" Dr. Stone was out of breath. His left side hurt and his breath was shallow and fast.

"I saw him as clearly as can be, standing there on his hind legs," Mike cried excitedly catching up with the others. "He did not look like a bear at all!"

"Of course...not. He's...no...bear. He's...Sasquatch." Dr. Stone's words came out one at a time. He was breathing hard, holding his side.

"Shouldn't you rest awhile, Dr. Stone?" Dave said.

"No. Let's go. We must get to this cave. We must photograph him."

They resumed their climb. A gray mountain, almost bare of vegetation, rose before them.

"How high up the mountain is the cave?" Dr. Stone stared at the treacherous steep upgrade.

"At least another two or three hundred yards, I think," Mike said.

"Let's go." Doggedly, Dr. Stone pushed ahead.

It was a hard climb. The shale covering the slope, was shifting under their feet, making them stumble and slip.

"Get going," Dr. Stone admonished them, seeing that Mike and Dave slowed down.

"Where does he get his energy?" Dave wondered in admiration.

The light was fading. The setting sun and the rising moon were simultaneously in the sky. Trying not to waste a moment of light, Dr. Stone desperately pushed himself up the mountain.

Suddenly, he stumbled. The loose shale slid from under his feet, making him fall backwards. With nothing to break his fall, he tumbled head over feet, down, down, down.

Mike and Dave gasped. They froze, but only for a moment. They rushed after the fallen man, slipping on the shale, sliding down, and getting up again.

Dr. Stone's fall was finally stopped by a boulder. He lay at its foot, unconscious.

Dave and Mike stared at him helplessly. His bruised body was grotesquely twisted. His head was bleeding, but the most terrible sight was his left leg. It was broken, the ragged bone protruding through his torn trousers. It was also bleeding profusely, the stones under him turning dark red.

"We must apply a tourniquet, or he'll bleed to death," Mike said taking his belt off.

Carefully, they tied the belt above the break. They bandaged Dr. Stone's leg, and then his head, using Dave's shirt, which he tore into strips.

"How are we going to get him down?"

"We must make a stretcher."

"From what?" The naked mountain could provide no material for making a stretcher.

"You stay with him. I'll go down to our camp and radio for a helicopter," Mike said.

"A helicopter won't come until morning."

"Do you have any better idea?" Dave had none.

"Okay. I'll be back with the blankets and food. Give me the camera. We'll have plenty to take care of without worrying about our cameras."

"Be careful."

"Sure. Take my jacket. Keep him warm, if you can."

"I hope he doesn't die on us," Dave said nervously.

Mike did not answer. He was on his way down, sliding and slipping, trying to keep his balance, three cameras swinging over his broad chest.

Dave covered Dr. Stone's motionless body with Mike's jacket, and prepared for the long night, his back propped against the boulder.

Several hours later, Mike returned, lighting his way with a powerful torch. He carried three blankets, crackers and dehydrated food and a can of Sterno in his backpack. Over his shoulder hung a canteen of water.

"The helicopter will be sent at dawn," he said, panting hard after the climb. "The paramedics told me not to move him nor try to revive him. They told me to loosen the tourniquet, though."

"He regained consciousness a couple of times," Dave said, helping to release the pressure of the tourniquet. "But he did not know what happened...and he didn't recognize me." They covered Dr. Stone with a blanket.

Mike lighted the can of Sterno and heated a small amount of water over its blue flame. They mixed a little of the dehydrated food powder with water and spread the mixture over the crackers. They did not talk. They sat wrapped in their blankets, chewing on their crackers, each thinking of the valiant old professor, lying helpless at their feet, his bones, as well as his dreams, broken.

They dozed off, and were awakened by the drone of a helicopter. They jumped up, waving their arms. The helicopter hovered over the mountain, getting ever lower, until it was only a few feet above their night encampment. A gondola, resembling a ski rescue team sled, was lowered from the open door of the helicopter. Two paramedics jumped down. They steadied themselves on the shifting shale, then proceeded to the prostrated form of Dr. Stone.

"Wow! A compound fracture of the tibia," one of them exclaimed.

"Is it very bad?"

"You bet. Especially since he probably has other injuries. Perhaps, even some internal injuries. But you did a good job with the tourniquet. You guys saved his life. He could've bled to death."

Gently, the men lifted Dr. Stone into the gondola. He moaned, but remained unconscious.

"We'll take him to the trauma center in Seattle," said the senior paramedic.

"Okay. See you in Seattle in a few days. Take good care of the old man. He's quite a guy," Mike and Dave shook hands with the paramedics. The gondola slowly began to creep up carrying the paramedics and the patient into the helicopter.

The machine rose vertically, then the pilot made a wide arc, and flew it toward Seattle.

Mike and Dave watched it getting smaller and smaller, becoming a tiny speck against the fiery sky of the early dawn, until it vanished beyond the mountains.

"Well," Dave said. "Let's go back and dismantle the camp. No more Sasquatch for us."

"Yeah. But we still have the pictures we took. What if they prove that we *did* come upon Sasquatch?"

Dave laughed. "Then we'll be famous!"

"*Rich* and famous," Mike corrected. "We'll be interviewed on television, and all the chicks at the campus will fight to go out with us!"

XIV

SASQUATCH!

Steve and Dan, having no knowledge of the termination of the expedition, resumed their journey as soon as the rain stopped.

The ground was inundated with water. Rivulets carved their way across the forest floor creating deep channels, twisting among the trees, uniting here and there into sizable streams. The forest stood still, weighted down by its dripping branches, surrounded by oppressive humidity.

Insects attacked the boys in force, sticking to their perspiring faces, climbing into their nostrils, drowning in the moisture of their eyes. Mosquitoes swarmed around them, stinging every exposed inch of their skins, covering them with itching, red welts.

"I bet you never thought that this beautiful forest could be so nasty," Dan exclaimed, slapping his forehead, and killing a mosquito bloated with blood. "By the way, we'll have to hunt for food. Sitting in the shack for three days, we finished all our supplies. Just as well. I'm tired of eating dry fish. I wouldn't mind having a piece of juicy meat. You wait here, I'll be right back." He lowered his backpack on a large flat rock blocking their path.

"Don't get lost. Without you I'll never find my way back."

"Don't worry. I'll be back in a jiffy." Dan checked his bow and arrows, secured the bowstring and tested its resilience. "Okay, I'm off!"

He was swallowed by the forest.

At once Steve felt uneasy. "What if something does happen to Dan?" he thought. "He could break a leg, or be attacked by a wild animal..." He watched anxiously the minute hand of his watch as it ticked the time away. Five minutes...ten...twenty.... The forest was mute except for the rustle of leaves and an occasional sharp cry of a bird. "I should've gone with him," Steve thought.

The sound of breaking twigs made Steve turn around. "At last!" he thought in relief.

Instead of Dan, he saw the hairy bulk of a large creature tearing through the bushes like a tank. It forged ahead, not more than fifty feet away.

"Sasquatch!" The thought streaked through Steve's mind. "My God! It's *Sasquatch!*"

The creature was gone. Steve yearned to rush in pursuit but his common sense prevailed. "No, I must wait for Dan. Dan!" he yelled, turning in all directions. "Dan! Dan!"

"What happened?" Dan appeared in the clearing. He carried a dead rabbit by its long ears.

"I think ... I'm *sure*, I saw Sasquatch!" Steve stammered.

Dan threw the rabbit on the rock and nocked a fresh arrow into his bow. The boys listened intently to the rustle of the forest, hoping to detect ominous sounds, but there were none.

"It's gone, whatever it was," Dan said. "But we must be careful. Whatever it was, it might return." Cautiously, stopping every few paces to listen, the boys crept forward until they reached the freshly broken branches.

"It left samples of its hair, look!" Dan said. Eagerly, Steve picked several tufts of brownish hair off the prickly branches and wrapped them in his handkerchief.

"What do you think it was?" he whispered, hoping that Dan would say, "Sasquatch."

"I don't know. The hair on the branches is at a height for a large bear...."

"It was Sasquatch!" Steve pointed to the ground, his face

suddenly gone pale. They boys stared at the giant footprint clearly visible on the soft surface of the ground. It looked like an imprint of a human foot with five toes and a broad heel.

"Just like the plaster cast I saw in a museum," Steve whispered. "How I wish we could make a cast of it! Let's at least photograph it." Steve focused his camera on the footprint. "Not enough light," he sighed, lowering the camera.

"Cheer up! We can protect the print. Let's make a marker. We'll pile up branches and stones around it.... Unless it rains again, the print might survive for a couple of days."

Steve brightened. "Of course! We'll bring Dr. Stone here. He will have all the necessary equipment for making a cast. Let's do it!" The boys quickly covered the footprint with hemlock and piled a small pyramid of stones over it.

"It's getting dark," Dan said. "Let's go back to the flat rock and set up camp for the night. We have a rabbit to cook!"

Steve was disappointed that Dan was not eager to resume their trek. But, he had to agree that it would be impossible to move through the forest in the dark. "Okay," he said trailing Dan back to their rock.

He observed, with fascination and revulsion, as Dan skinned the rabbit. Making several incisions with his hunting knife, Dan peeled the skin off in one piece, turning it inside out as if it were a stocking. He disemboweled the rabbit by swiftly slicing its belly open and scooping its intestines out with one motion of his hand. Then he wiped the cavity with leaves and, sticking a twig through the carcass, impaled it.

"Where did you learn to do that?"

"Any Native knows how to skin a rabbit! Big deal!"

They built a fire and spread their sleeping bags next to it. The fire kept flickering, the damp wood generating more smoke than heat. When the fire had finally produced enough glowing embers, Dan transferred them to a small pit that he had dug for the purpose. He suspended the rabbit carcass above the pit on two Y-shaped branches. Soon the aroma of roasting meat wafted through the air. Dan kept turning the skewer, allowing the meat to turn golden brown and drip with juices. "All we need now is a bottle of ketchup," he joked. He split the sizzling carcass into two portions with his knife.

"Just tell me if you have *ever* tasted anything more delicious than this!" he exclaimed, digging his teeth into the juicy meat.

"No, never! Never in my whole life!"

They trudged through the forest for two more days in search of Dr. Stone's camp. On the third day they came upon a rapids churning across their path.

"This is it," Dan said. "The final obstacle. We can either follow the course of the stream to where it flows into a river, or we can ford the rapids here and save a few hours."

"Let's ford the rapids. I'm eager to reach Dr. Stone."

"So am I. His camp ought to be at the junction. It's the most logical place for it. Let's go!" Dan tightened the straps of his backpack and jumped on the rock protruding from the water. Steve followed.

It was easier to cross the stream this time. Steve felt more sure of himself having crossed several small streams along the way. He joined Dan on a large rock in the middle of the fast-moving rapids.

"Take a picture of me." He handed his camera to Dan. "I have dozens of pictures of others but none of myself."

"Sure. Say 'cheese.'" Dan aimed the camera, stepping back for a better focus. It was a mistake. He lost his balance and plummeted down the rock into the swirling rapids.

"Dan!" Steve threw himself flat on the rock, trying to reach for his thrashing friend. The current pulled Dan away from the rock.

"Hold on to the stick," Steve shouted. He proffered Dan the stout sapling cane that he had cut for himself. Dan grabbed the stick. He struggled to stay afloat, his heavy pack dragging him down.

"Hang on, I'll get you out!" The raging rapids swirled around Dan, pulling the stick out of his hands, threatening to carry him downstream to a certain death. The boys fought with the rapids holding on to the opposite ends of the stick.

Steve inched his hands down along the stick until he touched Dan's hand. Instantly, he grabbed it. Both boys realized that the next few moments would be crucial. Steve had to lift Dan out of the water and away from its fatal pull. He had to drag him up, over the smooth rock which provided no purchase for his own body.

"Rest a moment," Dan coughed out. "Gather your strength.... I'm okay."

Steve relaxed his body without releasing his grip on Dan's wrist.

"Okay, here we go." He began to inch backwards over the bare surface of the rock, using his elbows for support.

He shifted to his knees as he slowly pulled Dan up to his waist. Dan helped to minimize his dead weight by seeking purchase on the rock with his free hand.

Slowly Steve pulled Dan all the way up. They both collapsed on the rock breathing heavily, exhausted by their terrifying ordeal.

"I lost your camera..."

"So what! You could've lost your life!"

Steve would not have been so cavalier about the loss of the camera had he known what was in store for him within a few hours.

The weather changed, becoming hot. They welcomed the change. Their clothes were still wet after their near tragedy at the rapids. They decided to camp in a little meadow on the shore. Dan stretched out in the tall grass, staring at the cloudless sky, his arms folded behind his head, waiting for his clothes to dry in the sun. Steve, sitting next to him, gorged himself on raspberries which he picked along the edges of the meadow. The air vibrated with the humming of millions of insects. Colorful butterflies fluttered from flower to flower, and birds chirped among the trees, filling the meadow with their joyful noise.

Suddenly, the birds took off in one twittering cloud, soaring into the sky. A hush fell upon the meadow.

Dan sat upright, tense, aware of the silence. Steve glanced at him in surprise, about to ask what was the matter as Dan mouthed the word *Sasquatch,* directing his gaze toward the edge of the meadow.

A hairy, towering, apelike creature stood there, bathed in full sunlight.

Sasquatch!

The creature was no more than thirty feet away from them, standing motionlessly.

Breathlessly, the boys watched, not daring to move.

Sasquatch stood quietly, like a tourist posing for pictures. Steve could clearly see his long arms reaching well below his knees, his bulging forehead, his flat nose with wide nostrils, and his protruding lower jaw.

Steve felt no fear of Sasquatch. He had dreamed of this moment so many times! And here it was, not a part of his fantasy, but real!

Sasquatch advanced a step toward them and began to feed on the berries. He plucked them one at a time as dexterously as an ape plucks a flea out of its mate's coat.

In his absorption with Sasquatch, Steve did not notice that Dan had stealthily reached for his bow and fitted it with an arrow. As Sasquatch advanced another pace toward them, Dan slowly raised his bow and aimed.

From the corner of his eye, Steve caught Dan's movement. "Danny! Don't!" he yelled as he fell upon Dan's body.

Steve's shout startled Sasquatch. He paused, and then charged back into the forest crashing through the thicket. The air was suddenly filled with the noxious odor, as if raw sulfurous gases had risen from the swamps.

Sasquatch was gone.

"You crazy idiot, you could have killed him!" Steve screamed furiously, pummeling Dan's back with his fists.

"Geez, I almost did," Dan did not defend himself. His lips were ashen.

Instantly disarmed by his friend's obvious remorse, Steve demanded in a calmer tone. "Why did you want to shoot?"

"I don't know. It was instinct. When I saw him moving closer, something in me took over. Thank goodness you stopped me. It would have been like killing a human being."

They stared at one another in disbelief, their hearts beating heavily, their faces covered with perspiration. They had just seen Sasquatch! Sasquatch, the dream of Dr. Stone and hundreds of others, the elusive creature whose existence no one was able to prove.

"If we only had your camera.... We could have taken several good snapshots of him!" Dan felt wretched.

Steve took hold of himself. "Do you think he will come back?"

"I doubt it. He's probably miles away by now. But he must've left his footprints in the sand. Let's look."

It was easy to see the spot where Sasquatch had crashed through the underbrush in his retreat. The branches were broken, and tufts of his reddish brown, coarse hair clung to the snapped twigs. Several deep impressions of footprints led straight into the creek.

"Well, here we are. We have the proof but we can't prove it!" Steve exclaimed, not hiding his desperation anymore. He felt tears welling up in his eyes.

"I'll never forgive myself for losing your camera!"

Steve was too crushed to say anything.

XV

Visit with Dr. Stone

In two more days, the boys finally reached Dr. Stone's base camp. It was deserted. The ashes around the fire pit were cold. "The camp is empty. Something must have happened," Steve said. "It must have become impossible to continue the expedition. The whole camp is dismantled."

Dan agreed. "We might as well head for home. It should be easier going back. We know the way."

Joey's dogs were the first to greet the boys. They rushed at them with happy yelps, prancing around them.

Joey appeared on the back porch. Behind him stood his ever-curious grandmother.

"Dad! The kids are back!"

"You grow two inch!" Grandma said looking up at Steve. "You grow big this way," she pointed to Steve's height, "and you grow small this way," she sucked her cheeks in, indicating that he had lost weight.

"What about me? Haven't I lost weight too?" Dan teased her.

"You Indian. Indian no lose weight. Indian gets tough!" She shuffled back into the kitchen, eager to feed the boys.

"Before you describe your adventures, I must tell you that

Dr. Stone already knows about the destruction of the lab," the Chief said leading the boys into the kitchen. "I had to inform him about it after his surgery." He briefly described Dr. Stone's misfortune.

"We knew that something very serious must have happened," Steve said. "Dr. Stone must have been picked up by a helicopter."

"Yes, he was. He's in a Seattle hospital, and he wants to see you."

"How did he take the news about the lab?" Steve guessed the answer.

"Very hard," Joey said. "He's terribly depressed. He even cried when Dad told him about the lab."

The Chief frowned. "Now, Joey, it's all hearsay."

"No it isn't. Lucy told Sandy that he cried and I heard Sandy tell it to Jim."

"I would cry too," Grandma said.

"Anyway," the Chief continued, "apparently Dr. Stone and his men saw some kind of a creature that they thought could have been Sasquatch. Mike and Dave brought along several very good plaster casts of giant footprints, which I admit, don't look like a bear's footprints. They also took a lot of pictures, and a video of some fuzzy creature running up the mountain slope. But it's hard to say what it was. It still could have been a giant bear."

Joey was impatiently waiting for his father to finish.

"Dr. Stone told Lucy that he was terribly disappointed in this expedition. Not only did he lose all his expensive equipment, thanks to Chickie, but his own search still brought no proof of Sasquatch. He said that there was no more money for his search. This year's expedition he had paid for out of his own pocket. So, he cried bitterly. He said that it was obviously not meant for him to be the discoverer of Sasquatch. He even said to Lucy that perhaps it was all fantasy, all in his head, and Sasquatch did not exist."

"Oh, but he does exist," Steve interrupted him. "Danny

and I saw him. Unfortunately, we, too, had an accident and lost our camera, so we couldn't photograph him. But we saw him clearly, in full sunlight, almost as close as we see you now. Sasquatch is real."

No one spoke. The Chief looked from Dan to Steve, reading their faces.

"Good Lord, the kids *did* see the creature! They're not lying!" he thought.

Cooing Dove turned away from her stove to look at the boys. "Steve Bradley, you make me proud. You one of us!"

Dr. Stone was dozing when the boys tapped at the half-open door of his hospital room.

"Come in," he called out, opening his eyes. "Ah, it's you, my friends. Come in, excuse me for not getting up, but, as you can see, I *can't* get up." He pointed to his leg encased in a plaster cast up to his waist. His head was bandaged also, and was supported by a thick collar, making his beard stick out as if it were not a part of his face, and had a life of his own.

Jim and the boys, feeling ill at ease in the hospital surroundings, crowded awkwardly around his bed.

"We need more chairs," Dr. Stone said. "Grab some from the hall."

"Well, tell me what have you been doing?" Dr. Stone said with artificial heartiness, beaming at them much too brightly. "But before you do, I want to establish certain rules," he reverted to an earnest tone of voice. "First, about my lab. It's gone. Kaput. Nothing can be done about it. Second, no commiseration. I have had enough of it already from my wife and my colleagues. And third, no talk about my future plans. I have none. Until I get out of here, I am making no plans. Understood?"

"Yes, sir," Joey said.

"Well, Steve, when are you leaving for home?" Dr. Stone pushed his glasses down his nose and peered at Steve over them.

"A week from Sunday. School is starting the following

week. Dr. Stone, I must tell you something important...." he began, blushing deeply.

"No, no, no!" Dr. Stone interrupted him impatiently. "I don't want to hear it!" In an exaggerated gesture of refusal to listen, he pressed his hands over his ears hidden under the bandages. "It was *not your fault* that the lab was destroyed. No explanations and no apologies, *ple-ease*! I heard it all from Lucy!"

"But that is not what I was going to tell you," Steve said, raising his voice.

Dr. Stone lowered his hands. "Well, all right, what did you want to tell me?"

"Dan and I, while searching for your camp, came upon Sasquatch. We *saw* him!"

Dr. Stone blinked. He tried to sit up but his cast would not allow him to. He inhaled deeply and held his breath. Finally, he hissed dramatically, "What did you say? You did what? You *saw* Sasquatch? Don't play tricks on me, boys!" he shouted, gesticulating wildly, nearly falling out of his bed.

"Take it easy, Dr. Stone, you'll hurt yourself," Jim said.

Dr. Stone ignored him. "Tell me, tell me about it!"

Steve and Dan, haltingly at first, began to recount their adventures in the mountains.

When they finished, Dr. Stone, a dejected smile on his lined face, remained silent for a long time. Finally, he said, "You are very fortunate, boys. Hundreds of people would give *anything* to see Sasquatch. I congratulate you. I must confess, I envy you. But obviously it was not meant for me to come up with the proof about Sasquatch." He chuckled bitterly. For a moment his mask of courage and good cheer slipped, and a face of disillusionment peered at the boys. It was the face of a disappointed man, and it stirred Steve's sympathy more than anything he had ever known. He felt tears welling up in his eyes.

"No, Dr. Stone! Don't say it!" he exclaimed hastily. "Dan and I know that if it hadn't been for your accident you would still be in the mountains right now. You would have found Sasquatch!

It looked as if he were on *your* trail as much as you were on *his*!"

Dr. Stone smiled thinly. "Thank you, my boy. Perhaps you're right. We had a feeling that Sasquatch was constantly watching us. We had plenty of signs of his presence. And then, when we pursued and almost cornered him, he slipped away from us. Our films are worthless. They're hazy, out of focus, and they prove absolutely nothing...anyway, under the circumstances, the best we can do now is to write down what we have seen. Can you describe every detail of your encounter with Sasquatch?"

"Sure we can," Steve said.

"Bring your notes to me as soon as possible. Make sure you keep copies. Perhaps we have enough material to write a book about our search for Sasquatch!"

"A book!"

"Why not? We've all had enough adventure to fill a three-hundred-page best seller! It will be considered fiction, of course, since we failed to come up with a proof, but it will still be a good read." Dr. Stone paused and smiled. "Start writing. Steve has only a few more days before he returns to California. I want your notes before he leaves. Shoo!" he clapped his hands.

Sheepishly, Jim and the boys left.

"Gee, a book? And I'll be one of the authors?" Dan shook his head in disbelief. "Cool!"

There were five postcards waiting for Steve when they returned to the village. His parents had written short notes from various places along the Greek coast where they were vacationing. They wrote that they missed him, worried whether his summer had been as good as theirs. He read the cards, smiling to himself. "Boy, if they only knew what a fantastic summer it has been!" he thought, arranging the cards in chronological order. Apparently they had been stuck in some dusty post office for weeks; they were all delivered at the same time.

He thought of the forthcoming Sunday when he would rejoin his family. He had never thought that he would miss his parents, yet here he was looking forward to going home; even though he was having such an exciting time among his new friends. In his mind's eye, he saw his youthful mother in her tennis dress with a green visor over her face to shade it from the sun. He saw his father, as always, at his computer, chewing the ends of his moustache in frustration when the story line failed to run smoothly.

"I never thought that I would miss my folks," he confessed to Joey as he showed him the postcards.

"I'm so glad that our dads were friends when they were young," Joey said. "Now you and I are friends. Let's *never* lose track of one another. And let's bring our dads together again!"

Dr. Stone phoned the next day. "I have been thinking all night about the book," he said. "If we are lucky and the book sells, it will provide us with enough money for a new expedition. In your report, boys, I want you to give me every detail, even the most minute."

"I remember when Sasquatch ran away, there was a tremendous stink in the air," Steve said.

"Describe it to me."

"Well," Steve hesitated, "It was...it sort of smelled of swamps. You know, it smelled like sulfur."

"Like rotten eggs," Dan prompted loudly.

"Yes. Yes, I know exactly what you mean. We smelled it too, when he came near our camp attracted by music." Dr. Stone's voice grew excited.

"But we did not smell it until he started running away," Dan was listening on the extension in the kitchen. "He must have been scared off by Steve's shout."

"Why did you shout?" Dr. Stone demanded angrily, as if Steve had betrayed him.

"Don't blame Steve. It was because of me," Dan hurried to explain. "I almost shot at him, and would have if Steve hadn't stopped me." Feeling miserable, Dan described how he had grabbed for his bow and arrows and how Steve's shout had saved Sasquatch and brought Dan back to his senses.

"Well, I'm mighty glad that you didn't release that arrow,"

Dr. Stone said, his voice becoming calmer. "Of course, there are many who would have liked to have Sasquatch's dead body as proof of his existence, but I am not one of them. Besides, you might have been accused of murder."

"Murder?"

"Yessir! In Scamania County there is an ordinance imposing a ten-thousand dollar fine for killing Bigfoot, in addition to the charge of murder. Did you know that?"

"No, sir."

"Well, my boy, it's lucky that you did not shoot him. Continue."

"Sasquatch showed no fear of us. I had a feeling that he was almost a timid creature," Steve said.

"Not me. I expected him to charge at us! It was like meeting with a great bear. One never knows what to expect of him," Dan explained.

"I tend to agree with you," Dr. Stone said. "Any wild creature met in his natural habitat may be quite dangerous. Anyway, write it all down. Describe *everything* that you saw. I'm sure we will have interesting material for a good adventure yarn!"

Two days later the Chief took the boys to Seattle to see Dr. Stone again and to deliver their notes.

Dr. Stone was not in his room. He was taken to the x-ray department and the floor nurse could not tell how long he would be gone.

"We'll call him later," the Chief said. "We'll leave these papers for him." They left their notes sealed in a manila envelope propped against Dr. Stone's pillow.

At the elevator they collided with Lucy. "I'm on my way upstairs to take dictation from Dr. Stone," she said. "Isn't it terrific? I mean, if it hadn't been for Chickie ransacking the lab you wouldn't have gone to the mountains and seen Sasquatch! If it were not for Chickie, Dr. Stone would not have come up with the idea of writing an *adventure* story!"

They laughed. "Some logic," Joey said sarcastically, but his father shot him a warning glance and Joey said no more.

"Well, say hello to Dr. Stone for us."

"Bye-bye." Lucy stepped into the elevator.

"I'll be leaving next week," Steve said

"I'll send you the invitation to my wedding," Lucy smiled as the elevator door closed.

XVI

Raven's Cousin

The last few days of Steve's vacation were coming to an end. He spent them with Joey and Dan, visiting their favorite haunts and shooting at targets on the beach.

"Don't forget, you guys are coming to California for Christmas," he kept saying. "We'll go to Disneyland and to the Universal Studios to see how the movies are made! My dad will get us passes. We'll have lots of fun!"

The night before his departure the Chief called Steve into his study. "You have been invited to a tribal meeting tonight," he said.

"It is a special honor for you," Cooing Dove said, entering the room.

"Grandma! Don't talk so much." Joey made a face at her. Grandma, Sandy and Jim had a conspiratorial air about them as they all gathered for supper at the Chief's house. They kept glancing at Steve, smiling enigmatically as if hiding some secret.

"What's with you people?" Steve finally said. "Why do you look at me like that?"

"You find out soon," Grandma cackled with delight. Joey shushed her again. Even the Chief had a mischievous smile on his face as his eyes met Steve's.

The supper over, all together they crossed the meadow toward the bonfire already burning on the grounds of the tribal council. The space around the fire was crowded with villagers, many of them wearing their native costumes. They greeted the Chief and his family warmly.

The Chief took his place among the elders, and faced the crowd.

"My fellow tribesmen," he began rising to his feet. "A proposal was presented to me the other day to bestow the honorary title of Raven's Cousin upon our guest from California, Steve Bradley."

Steve felt blood rush to his face, turning his ears crimson. "Good God! What *is* it?" he panicked.

"You all know Steve Bradley by now. He is the one who kept us from losing the archery competitions, and, he is the one who saved the life of Danny Beaver Tooth."

There was a murmur among the villagers.

"Yes, my friends. You know that Steve Bradley and Danny Beaver Tooth went into the mountains in search of Dr. Stone after his equipment was destroyed by one of our own people, my own nephew, to my everlasting shame. Well, Steve proved that he was a good match for any of our people. Dan Beaver Tooth reported to me that Steve was as fearless and dependable on the trail. When Dan fell into the rapids, Steve did not panic. He courageously fought the river and saved Danny's life. But this is not all. Together the boys met with Sasquatch!" The Chief paused and puffed on his pipe. Realizing that it was empty, he stuck it into his pocket and continued. "As you know, I have always been skeptical about Sasquatch. I have never seen one. But many people have claimed to have seen him; however, usually from a distance or in the dark. The boys swear that they saw him in broad daylight within about thirty feet. They could not take any pictures, unfortunately, because due to an accident which almost cost our Danny his life, they lost their camera. Never mind. There's always another year, as our friend Dr. Stone used to say. Anyway, the Council has decided to recommend that Steven Bradley be elected an honorary member of the Raven Clan!" The Chief peered at his tribesmen from under his thick eyebrows, looking like a sleek, handsome raven himself.

"Anyone opposed? No. The proposal is accepted."

The villagers applauded. Steve watched them as they sat cross-legged around the fire, their faces immobile, as if carved of dark mahogany. A wave of affection toward them swept over him.

"Steve Bradley, please approach the Council," the Chief said.

Steve's feet felt wooden as he stood up. Awkwardly he walked toward the Chief and the elders.

"Steve Bradley, we proclaim you the Honorary Cousin of the Raven Clan. From now on you are the cousin of every member of the clan. Jim, the headdress."

Jim reached under his poncho and took out a leather headband decorated with beads and crowned with a single bluish-black raven's feather.

The Chief brushed Steve's blond hair off his forehead. Then he placed the headband over Steve's head.

"It's only the third time that our clan has bestowed such an honor, and the first time *ever* on a white man. Be proud of this honor." The Chief shook Steve's hand.

The crowd was waiting for Steve's reply. He blushed deeply, feeling painful shyness.

"Ladies and gentlemen!" He hesitated. "No, if I may—my dear cousins," he corrected himself.

The villagers loved it. "Right on!" Joey shouted.

Encouraged, Steve continued, "I am deeply honored to be named an Honorary Cousin. I don't know what my duties are, but I'm sure my cousins Joey Little Raven and Danny Beaver Tooth will teach me." He bowed awkwardly from the waist and not knowing what else to say turned back to his seat.

The Chief stopped him. "Not so fast, Raven's Cousin. The shaman must perform the initiation rites."

A sick sensation of fear grabbed Steve at the pit of his

stomach. The shaman! What would he do to him? Would he draw blood? Would he make him perform some disagreeable task?

Steve heard the sound of the shaman's rattles. He swallowed hard, preparing himself for the worst as waves of fear rose from the pit of his stomach.

The witch doctor sprung out of the darkness beyond the circle of bonfire light. Has wiry body shone from the applications of halibut oil, and his mane of matted hair bounced over his bony shoulders.

"You brave boy, bar-bethld, white man. I proud, to-to-bush. You good archer. You become Raven, klook-shood. You be good klook-shood."

"Repeat after him," Jim said softly.

"I'll be a good klook-shood," Steve repeated. "Such a silly sounding word," he thought, fighting an urge to laugh. He tried to keep a straight face. He remembered Joey's warning to take the ceremonies seriously. "We have great pride in our rituals," Joey had explained to him some weeks earlier preparing him for the potlatch. "In the old days if a person made fun of our ceremonies his lips would be skewered right through the flesh with sticks, just like a stuffed turkey ready for roasting!"

Steve made an effort to retain a serious expression. He knew that this savage practice had been abandoned long ago, but perhaps there were other, less drastic forms of punishment.

The shaman's breath felt hot on Steve's face. He began to smear red and black paint on Steve's cheeks and forehead, dancing around him, pointing with his rattle to Steve's hair—"Appsahp kle-sook!"

"He says that you have white hair," Jim translated in an undertone.

The shaman whirled faster, shaking his rattles with one hand, beating the drum with the other, like a busy one-man band. Suddenly he stopped, freezing on one foot like a stork and then collapsed on the ground in a heap of dirty rags and matted hair. His eyes rolled up under his lids until only the whites were visible. He shouted, "Klook-shood kle-sook to-to-bush!" as the foamy saliva gathered at the corners of his mouth. He was in a trance.

"What did he say?" Steve felt frightened, shaken by the shaman's performance.

Cooing Dove shuffled forward and pushed Jim aside. She carried her precious Chilkat blanket over her arm.

"He said that he's proud of the White Raven. You're in, kid. You're Raven's Cousin. Congratulations!" Jim said, slapping Steve on his back.

"For you to take home." Grandma handed Steve the blanket. "Belonged to Big Chief, my father." She helped Steve to drape the blanket over his shoulders.

"Thanks, Grandma, I'll treasure it always," Steve embraced the old woman, and kissed her soundly on her withered cheeks.

"Like on TV!" she cackled.

"And this is from us," Sandy, the Indian goddess of his dreams, stood smiling at him, an antelope-skin vest embroidered with tiny beads in her hands.

"Jim made the vest, and I decorated it," she said. "We deliberately made it larger so you would have more years to wear it."

"Oh, Sandy...you're so beautiful...I love you..." Steve thought, not seeing the vest but only her face. He blushed deeply under the garish paints, glad that no one, least of all Sandy, had noticed his lapse.

She handed him the vest, and touched his cheek with her lips lightly.

"Gee, thanks..." he stammered, his cheek burning where she kissed it.

The villagers surrounded him now, offering their congratulations. Many brought him small gifts such as baskets, wood carvings, and necklaces made of seashells.

"Now you must dance. Make your clan happy," Cooing Dove said, clapping her hands.

A momentary panic seized him. Dance? He couldn't dance. But then he recalled the ceremonial dance of introduction during the potlatch. He grinned at Grandma. "Just watch me! It's for you!"

He stepped into the bright spot before the fire and began to chant "Klook-shood, klook-shood, raven, raven." He made three little steps, turning around and freezing, displaying to the crowd his priceless Chilkat blanket.

Sandy, and then Jim, joined him, followed by Joey and Dan and the other villagers, one by one. The witch doctor's apprentice began to beat his drum rhythmically as the Raven's Clan celebrated the investiture of its only white honorary cousin.

XVII

Going Home

In the morning Steve was to leave. He slept badly during the night, waking up often. He regretted that he had to return to his mundane life of school, tennis and French lessons. But then he would think of his parents, and how much they would enjoy hearing about his adventures.

But most of all his thoughts were of Sasquatch. In his mind he pictured him the way he had appeared that day, deep in the mountain wilderness. He thought of Dr. Stone and of his deep disappointment that once again Sasquatch had eluded him. "I sure hope that his book will be a smash hit. I know that he will return next year and resume his search. Perhaps he'll allow Dan and Joey and me to join him next time. Boy, that would be cool!"

Then, he thought of Lucy and Sandy, still in love with both of them, wondering whether he would ever meet such wonderful young women when *he* would be old enough to date. And, he thought of Chickie. "Poor guy, locked up in some juvenile detention center probably hating *the whole world* by now, not just me," he thought. "But if it hadn't been for him, as Lucy said, I wouldn't have seen Sasquatch."

He tossed wakefully, finally falling asleep, only to be awakened by Grandma a short time later. It was time to go.

Cooing Dove had already packed his mementos, and his

bow and arrows, the gift of the Nootka Chief. "I always pack for my son. I pack for you," she said.

Seated at the table in her kitchen, seeing her move slowly around the stove, Steve felt a glow of tenderness toward her. "I might never see her again," he thought. "How old is she? Eighty? Eighty-five? I love her!"

They ate breakfast in almost total silence, Joey and Steve both depressed over their forthcoming separation. The Chief pretended to be reading but he too, often glanced at Steve, loath to see him leave. He came to love Steve as one of his own.

The whole populace of the reservation gathered in front of the Chief's house. Everyone wanted to go to the airport to see Steve off.

The street was congested with cars and trucks of every make and vintage. The village's only policeman frantically organized them into a semblance of a motorcade.

People were dressed in their best clothes, children running between the vehicles, unable to control their excitement about going to the airport. Many adults as well had never been to an airport. They looked forward to escorting Steve, their Honorary Cousin, to the plane.

A school bus pulled up and the policemen directed a group of elderly men and women to take their places on the bus. Like everyone else, they wished to go to the airport.

The Chief drove the leading car. The boys, sitting together in the back seat, were subdued by the sadness of their parting. But the passengers of the other vehicles in the convoy were having the time of their lives. The younger men honked their horns, played radios at top volume and drag raced one another along the wider stretches of the road. But they stayed well behind the lead car. They all knew the peril of provoking the Chief's anger.

The travelers became aware of the traffic as soon as they left the country roads behind them. The Chief halted the convoy and directed Jim to send a message to the highway patrol over his CB radio, requesting an escort. Shortly, two patrolmen on motorcycles, their lights flashing, joined the group. They pulled up in front of the motorcade, leading it into the traffic and pacing it at an even fifty miles per hour.

The boys' gloomy mood changed. It was exciting to be escorted by the motorcycle patrolmen. "We're big shots!" Joey whispered puffing his cheeks out, pretending to be a fat man. The boys giggled.

The Seattle-Tacoma airport was still way ahead but its existence was apparent long before the motorcade had reached it. Both sides of the highway became congested with warehouses, hangars and large structures bearing the names of domestic and international airlines and air freight companies. The skies overhead had suddenly become full of the screech of jet engines and the whirl of helicopters.

The patrolmen led the convoy to the ramp and then on to the airport, escorting them to a special VIP parking lot.

There were only forty minutes left before the departure.

"We sure don't have much time for saying good-bye," Joey said.

"It's better that way." Cooing Dove was, as always, practical.

"Well, my boy, fare thee well," The Chief embraced Steve. "It was a great pleasure to have you among us. I don't have to tell you that you're always welcome to return. Remember that!"

"We must hurry," Jim reminded. They moved toward the terminal at a run, one huge, exuberant crowd.

"Take the people to the observation deck," the Chief ordered.

Steve shook hands all over again as one family after another surrounded him, passing him on to the next group. It seemed that no one had any concept of airline schedules.

At last Jim separated Steve from his well-wishers. He led the group to the observation deck. Only the Chief, Joey and Dan, Cooing Dove and Sandy remained with Steve, following him at a half-run to the check-in counter and then to the last checkpoint.

"Western Airlines, the last call for Flight 626 to Portland, San Francisco and Los Angeles. All aboard, please," the voice intoned over the public address system.

"That means me," Steve said. "I must go!" Cooing Dove affectionately ruffled his hair. Joey and Dan slapped him on his back. Sandy kissed him.

"I'll be back...I'll be back!" Steve kept repeating, feeling himself near tears from all this outpouring of affection. "I'll be back, I promise!"

Steve tore himself away from his friends. He walked through a metal detector, collecting his flight bag as it slid toward him on a conveyer belt. He waved to his friends once more, and disappeared in a narrow corridor leading to the plane.

A flight attendant, smiling, greeted him with professional courtesy.

"Ah, you must be *the* Master Bradley for whom the departure of the aircraft was delayed," she said. "May I show you to your seat?"

Steve settled into his seat and fastened his safety belt. He peered at the terminal building through the small window next to his seat and laughed.

The observation deck was teeming with the colorful crowd. But what had caught Steve's eye was a towering figure in a gorilla suit holding a placard. The huge lettering was beautifully stenciled in black and red colors and it read: HURRY BACK RAVEN'S COUSIN—YOUR FRIEND SASQUATCH.

"Is this demonstration in your honor?" the flight attendant asked. "What did you say your name was?"

"Steven Bradley, Raven's Cousin," Steve replied proudly.